Disruption

Emerging Technologies and the Future of Work

Victor del Rosal

Emtechub
Dublin

Published by Emtechub, Dublin
www.emtechub.com/disruption

Publication Date: September 12, 2015

Author photograph by Michael McLaughlin

Sold by: Amazon

ISBN-13: 978-1514173947
ISBN-10: 1514173948

BISAC: Technology & Engineering / General

For Laura, Lucía and Alexander

Science, my lad, is made up of mistakes, but they are mistakes which it is useful to make, because they lead little by little to the truth.

–Jules Verne

Thank you

To my family, for enduring months of hearing: "in the future…"

To Dr. Luis Flores Castillo (member of the team that discovered the Higgs boson at CERN) for taking the time to painstakingly read the original manuscript and make significant contributions and edits; my heartfelt gratitude

To the wonderful Advisors of the Emerging Technology Leaders Global Initiative and the young Emtechleaders Class of 2016

To Mayo County Enterprise Board (now LEO Mayo) for being early supporters of the Emtechub endeavor

To John Magee, for his thoughtful and thorough review; my most sincere thanks

To MexNet and Red Global MX, for serving as inspiration

And a special thank you to young scientists- and technologists-in-the-making… may these pages stir your imagination.

About the Author

Victor del Rosal

Founder of Emtechub, emerging technology education startup

Lecturer at the School of Computing of National College of Ireland, Dublin

Masters in Science (Honors) from UCD Michael Smurfit Graduate Business School, Dublin, Ireland

Diploma in Economics, Harvard University, Cambridge, MA, USA

Bachelor's Degree in Industrial and Systems Engineering Monterrey Tech, Monterrey, Mexico

Director of Strategy at CloudStrong Ltd., Castlebar, Ireland

Founding President of MexNet, the Irish chapter of Red Global MX, Dublin, Ireland

Head of Business Analysis at the Supply Chain Center of Excellence, TCS, Dublin, Ireland

Managing Director, W8 Training and Consulting firm, USA-Mexico

Co-founder and CEO of the Prince of Wales's Youth Business initiative, Mexico City

Consultant to the Inter-American Development Bank, Washington, DC, USA

World Bank scholarship for internship at the Boston-based Youth Employment Summit/ Education Development Center think tank in Boston, MA, USA

Contents

Thank you.. ix

About the Author.. xi

Foreword... xv

Why *Disruption*?.. xix

PART I

Emerging Technologies: Enablers of Disruption

1. Disrupted future 1

2. Artificial Intelligence 7

3. Autonomous Vehicles 19

4. Flying drones... 27

5. 3D Printing 33

6. Internet of Things... 41

7. Virtual and Augmented Reality........................... 47

8. Biotechnology.. 55

9. Alternative Energies 61

10. Blockchain 65

PART II

The Future of Work: Shifts Driven by Emerging Technologies

11. Reality blurred 73

12. Blackbox combinations.....................................79

13. Augmented humanity 83

14. Towards full automation 89

PART III

Getting Ready for Disruption: Preparing for Exponential Technological Change

15. Education for disruptors..................................103

16. Solving problems...123

17. Engineering security 129

18. The transition to prosperity............................. 137

Conclusions and final thoughts............................. 141

Keeping in touch 149

Foreword

Innovation is not something that can be produced on-demand—there are no "recipes" for crafting technological breakthroughs. Instead we can aim at increasing the probability of triggering the kind of associations (among people or among ideas) that end up producing disruptive technologies. Although this is also a tall order, there are a few examples of organizations that seem to fit the bill.

CERN

One of these is the European Organization for Nuclear Research (CERN). Its mere size is impressive: a total of about ten thousand scientists are associated with it. It hosts the largest and most powerful particle accelerator ever created, the famous Large Hadron Collider, installed inside an underground tunnel that describes a 27-kilometer nearly-circular path one hundred meters below the French-Swiss border.

Never before had such a large scientific collaboration been formed, neither had a single machine of this magnitude been built. Along this monstrous machine, protons are accelerated to almost the speed of light in opposite directions, to make them collide in the center of four gigantic particle detectors, each rivaling the size of a medieval cathedral.

Creators of disruptive technologies

It is no coincidence that this organization—concurrently so large and ambitious—is at the forefront of science and technology, and an early adopter of disruptive technologies—some of which have even been developed here.

As a relatively well-known example, the need for an effective and fast communication tool for the dissemination and discussion of scientific results led to the development of nothing less than *the Internet*. Of course, there is no need to dwell on the relevance and transformative power of this development.

To illustrate the relatively early adoption of disruptive technologies, many research teams at CERN utilize remote meeting software to collaborate effectively and continuously from locations in all continents; a professor from a Chinese university may have frequent meetings with students and postdoctoral researchers, some in Beijing and some stationed at CERN (in Switzerland), while he visits an institute in the US. This practice has been in effect for several years already. Face-to-face video interaction, screen sharing, chat box exchanges, and easy file sharing sometimes make it preferable to hosting a meeting in this fashion than walking into a meeting room in the building next door.

Pursuing ambitious scientific goals at CERN and similar laboratories has prompted several important technological developments. The need to analyze the massive amounts of data produced in these laboratories led to an accelerated development of computing clusters, which later evolved into what we now call *cloud computing*. Multivariate methods developed originally for these analyses have been adopted for *data mining*.

The strong investment in the development of superconducting magnets (used to steer the beams of particles) has had a strong impact not only in the magnet industry but, through it, in the nuclear magnetic resonance industry.

Particle accelerators have also allowed the development of cancer therapies that use protons or carbon ions instead of radiation, which can better target tumor tissue. These technologies, along with several others, are in constant research and development at these laboratories.

Disruption

Now more than ever, the rate of technological innovation maintains its ever-faster march, without any concession to our inability to keep up with it.

In *Disruption*, Victor del Rosal has put together a comprehensive and coherent list of those developments that are likely to redefine, in the next few years, the way in which we live and work.

Whether or not you have kept up with the mind-blowing pace of these emergent game-changers, I think you are likely to enjoy *Disruption* as an invitation to think about the magnitude of the changes to expect in the near future, as a starting point of more research on several topics, and—hopefully—as a trigger that may lead to new disruptive uses of these technologies.

Dr. Luis Flores Castillo
Member of the team that discovered the Higgs boson
CERN, Geneva, Switzerland
September, 2015

Why *Disruption?*

AN INTRODUCTION

Disruption: verb \dis-ˈrəpt\ *to cause (something) to be unable to continue in the normal way; to interrupt the normal progress or activity of (something).* –*Merriam Webster dictionary.*

I wrote this book to satisfy my own curiosity. I wanted to know where technology is heading in the next ten years or so, because, among other things I have a young daughter who will certainly start asking all sorts of questions—she already is.

The journey started four years ago, when I was working for a management consulting firm in the telecom space. I needed to know which trends and technologies could impact our clients, most of them *Fortune 500* companies, exhibiting the greatest growth and potential. I was hooked.

To say that we are on the verge of transformation driven by exponential technologies is quite an understatement.

Note that, for the purpose of this book when referring to technologies we will use the adjective *exponential* and *emerging* interchangeably. These refer to technologies with the potential to cause disruptive changes across the board.

Speed of innovation

One of the unforeseen challenges of writing a book about emerging technologies is the speed at which they move. Consider a recent announcement like this one: **Uber to buy 500,000 autonomous vehicles from Tesla**. The surprising thing is that statements like this one seem to pop-up in the news on a weekly basis. Innovation cycles are clearly accelerating.

A book about humans

This is a book about technology but more importantly, this is a book about us—to help understand the role of humans in an increasingly technological society.

I think we have a moral obligation, especially with the generations following us, to provide some sort of guidance. We are entering territories which are exciting and unchartered.

Structure

The book is divided into three parts. In the first one we explore emerging technologies which will likely play a significant role in the future of work. The idea is to introduce some of the building blocks which could enable significant leaps across industries and society.

In the second part we explore the shifts which are likely to occur driven by these technologies, answering questions like: How could technology change the way we work? What are the patterns in the occupations which are likely to be in high demand and those which will probably fade? What other major work-related trends could be significant?

In the final part we will look at some ideas for preparing for the future of work.

An overview of emerging technologies

Disruption is intended to serve as a *primer* to pique your interest in tech fronts you may have not previously considered. You will gain a glimpse into some of the emerging technologies with greatest disruptive potential. Perhaps more importantly, you may extract insights that may be useful regardless of the state of any given technology.

We will explore things like drones, driverless taxis, the virtual reality office, the Internet of Things, among other technologies, and how all this could change how we work.

Automatic fulfillment of groceries

Your fridge senses that you are about to run out of milk. It automatically triggers a process to ensure you are well stocked: it sends a message to your local grocery store. There, a team of robots assemble your order, automatically debiting your account. A small box with your order is fitted onto a drone capable of transporting a payload of up to five kilograms. The drone goes to the store's departure area and takes off. Within ten minutes your order has safely arrived to your house. It is received by your robot concierge—something resembling a vacuum cleaner. It opens the door, retrieves the shipment, and stores the delivery inside your fridge. Is this science fiction or science fact?

No fad, no hype

Just as the Internet could have been dismissed as a passing fad in the mid-1990s, there are powerful emerging technologies brewing that could easily be deemed as irrelevant to our future. However, if history has taught us something, it is that we cannot afford to discount that which we do not understand.

We are at the brink of a major technological and labor shakeup, stemming from overlapping innovation and technology advancing at an exponential pace. Hence, changes will be felt not in 20 years but in a window of five to ten years.

Big changes underway

Imagine being at the onset of the Internet revolution around 1995 and conjuring images of companies such as Google, Facebook, Skype, WhatsApp, Airbnb, Uber.

While it is very difficult to make predictions, especially about the future—paraphrasing physicist Neils Bohr—one thing we can be certain of: we are at a unique time in history powered by exponential technologies. This enables disruption at a truly unprecedented pace, presenting us with both opportunities and growing pains.

It is easy to overlook the significance—and disruptive consequences—of the fact that in the next ten years we will *not* experience the same magnitude of change we have experienced in the previous ten; it will be more like decades of innovation, compressed into a shorter frame. The recent IT revolution serves us a warning of how much the economy can change in such little time. The world may be so deeply transformed that by 2025, in retrospect, the Internet revolution may look like a warmup.

Victor del Rosal

July 2015

PART I

Emerging Technologies:
Enablers of Disruption

1

Disrupted future

EXPONENTIAL COMPUTING POWER GROWTH

Any sufficiently advanced technology is equivalent to magic.
—Sir Arthur C. Clarke

G o back in time to the mid-1990s. You have just received your first email, connected via a dial-up modem with a top speed of 56 *kilobytes* per second (that is 1/18th of a megabyte per second). You hear about this thing called the *World Wide Web*. You are thinking: *Is this a fad? Is it a just a geeky way to communicate? Will it be actually useful? Will it ever go mainstream?*

In 1995 email barely existed (Hotmail launched in 1996), faxing documents was still the norm, there were no smartphones, no Wi-Fi, Google would not be invented for *three more years* (1998), and social media was limited to text-based bulletin board systems. The idea of making a video phone call from a handheld device—never mind for free—was unimaginable. It was not easy to fathom the degree to which the Internet would transform the world, serving as a platform to spark innovation.

But what if four, five, or more technologies as revolutionary as the Internet were about to mature within the next decade? What would be the disruptive power of a few Internets *combined,* going mainstream within a shorter time span? How would this reshape the economy?

This is where we stand today.

Exponential growth in computing power

In 1965, the co-founder of Intel, Gordon Moore, observed that the number of transistors per square inch on integrated circuits had doubled every year since its invention. Moore predicted that this trend would continue for the foreseeable future. Over time this became known as Moore's Law, and the doubling of computing power was corrected to every eighteen months. This has remained constant over the past five decades.

In July 2015, IBM announced a 7-nanometer chip, breaking the difficult 10-nanometer barrier (a human hair is approximately 90,000 nanometers thick and a strand of human DNA is 2.5 nanometers thick) and paving the way for the continued exponential growth of computing power—Moore's Law is alive and well.

MIT's Technology Review points out that "even smaller computing devices will proliferate, paving the way for new mobile computing and communications applications that vastly increase our ability to collect and use data in real time."

The consequence of exponential growth in computing power is that in our pockets we carry (and take for granted) devices which are *as powerful* as the military-grade supercomputers of the late 1980s and early 1990s. Unlike them, however, our smartphones do not require nitrogen cooling and do not take up the space of a bedroom!

Moreover, by virtue of Moore's Law, by 2025 your smartphone will be about *30 times* more powerful than your existing one. What could you do today if you had the power of 30 smartphones crammed into a single device? This is happening within 10 years. And to start painting a picture of how our work life could shift by 2025, let's start by looking at our most immediate computing devices.

The future of the smartphone

By 2025 your smartphone will have become your main computing device. However it will not be what we expect. It will be more like a mix and match of technologies and connected devices. With it you can comfortably fulfill virtually all your business and personal needs. Laptops, desktops, and larger devices will become largely redundant, confined to niche applications. The chief researcher at Kantar market research and Ex-Gartner research VP, Carolina Milanesi, agrees.

As the price of computing power keeps dropping, devices become commodities, becoming increasingly accessible to larger sectors of the population. The sheer increase in computing power will have profound effects in society as will be explored in other chapters. This democratization of access is important as it continues to level the playing field.

Interacting in novel ways

Your personal device will be even more interconnected with other devices, eliminating the need for an inbuilt smartphone screen. Instead, you will be able to project visuals onto virtually anything. Your eye wear (prescription glasses, contact lenses, etc.) will become your screen, thanks to advances in augmented reality (AR), allowing for text and graphics to be superimposed on anything you look at, including everyday surfaces such as walls. Companies including Google (with Google Glass as an early attempt) are notable AR players. A company called Innovega has been working a number of years on ioptik, a contact lens onto which you can project images, replacing larger eyewear. A Japanese company, Aerial Burton, has devised touchable 3D holograms, using super-fast femtosecond lasers. Their device can show text and pictures in mid-air, without using a screen.

Advances in natural motion sensing and tracking, including hand gestures, eyeball tracking, haptic touch, etc. will increase our capabilities to input information, increasing the wealth of expressions that can be captured. Companies including Leap Motion, and virtual reality players including Oculus Rift, HTC-Valve, Samsung, among many others are making progress in this area.

PrimeSense, a company bought by Apple, pioneered the concept of 3D sensing, which allows for displays to be manipulating by gestures, much like what the Kinect does, Microsoft motion sensing input devices used for sensing motion in the Xbox. Shel Israel, Forbes contributor explains that, "essentially, [this technology] lets machines see you or your general location in 3D, allowing the machines to learn the context of a situation. Its sensors are so good they can actually see the heartbeat of game players, thus understanding how much a player is enjoying a game or not."

Ubiquitous Internet

As a connectivity backbone, we will need super-fast and reliable Internet that simply works. Uninterrupted Internet access will be accepted as a human right, accessible regardless of geography, location, provider, etc. Connectivity will be virtually guaranteed for everyone everywhere on Earth.

A study by Pew Research Center highlights that by 2025 the Internet will be everywhere, it will become ubiquitous, just as electricity is today. Joe Touch, director of the University of Southern California's

Information Sciences Institute points out that "We won't think about 'going online' or 'looking on the Internet' for something. We'll just be online, and just look."

Initiatives such as Internet.org by Facebook, Project Loon by Google, OneWeb by Virgin are already working in this direction. This will mean that global Internet population will jump more than twofold from three to eight billion. Singularity University founder Peter Diamandis points out that "the additional five billion Internet users represent tens of trillions of new dollars flowing into the global economy."

Increasingly powerful applications, services, gaming, etc. will be "streamed" from the cloud. For instance, take IBM's Watson—the artificial intelligence computer which defeated the *human* world champion of *Jeopardy!* IBM has made Watson available as a cloud service.

Watson-as-a-Service means that an AI which would be otherwise completely out of reach can now be your new AI agent or a "freelancer" on steroids, an outsourced service. This is profound: a cloud service that you plug into your small business or health care organization to increase your ability to provide customer service or to identify viable treatment options for cancer patients, powered by a supercomputer as an ally. Did we even dream of this in 1995?

2

Artificial Intelligence
THE BASIS FOR AUTOMATION

"We are on the edge of change comparable to the rise of human life on Earth." —Vernor Vinge

Artificial intelligence (AI) has been the focus of science fiction writers and film producers for decades. Contrary to Hollywood's typical portrayal of AI as an evil overlord, the *weak* type of AI already pervades the modern world in many useful ways.

What is Artificial Intelligence?

Artificial intelligence (AI) is the branch of computer science concerned with making computers behave like humans. Major AI researchers and textbooks define this field as "the study and design of intelligent agents", in which an intelligent agent is a system that perceives its environment and takes actions that maximize its chances of success. MIT's John McCarthy, who created the term in 1955, defines it as "the science and engineering of making intelligent machines".

One way to classify AI is in relation to human intelligence, that is, into three categories: narrow, general, and super intelligence.

Artificial Narrow Intelligence (ANI)

Also known as weak AI, it is a type of artificial intelligence that, as the name suggests, is narrowly focused on one type of work.

It turns out we are already very familiar with this type of AI. From a consumer level, you can see Artificial Narrow Intelligence (ANI), for example, in the way your bank detects abnormal spending patterns, with algorithms specialized at recognizing suspicious transactions. Apple's Siri and Google Now virtual assistants are also a type of ANI. The recommendations given by Amazon, Netflix, and LinkedIn, tailored to suit your particular preferences, connections, and so forth, are also a type of ANI, based on highly refined algorithms.

Two of the most public demonstrations of narrow AI are IBM Deep Blue's defeat of world chess champion Garry Kasparov in 1995 and the defeat of the champion of *Jeopardy!* by IBM's Watson. Tim Urban, editor of *Wait But Why*, points out that "the world's best Checkers, Chess, Scrabble, Backgammon, and Othello players are now all ANI systems."

Today more than half of equity shares traded on US markets are performed by bots, that is, algorithmic high-frequency ANI agents that respond in milliseconds to complete trades that match parameters set by investors.

Artificial General Intelligence (AGI)

While ANI agents may be impressive doing highly specialized work, they are no match for the *combined* range of complex functions performed by human beings.

Artificial General Intelligence is the type of artificial intelligence that could match human intelligence, being able to perform any intellectual task that is humanly possible.

This includes the ability to "reason, plan, solve problems, think abstractly, comprehend complex ideas, learn quickly, and learn from experience," as described by Professor Linda Gottfredson.

To this day, AGI has not been achieved. How can we know when it is realized? Luke Muehlhauser, of the Machine Intelligence Research Institute presents four tests for AGI which have been proposed.

Turing Test

Perhaps the most well-known one is the Turing Test, examining a machine's ability to exhibit intelligent behavior equivalent to, or indistinguishable from, that of a human. Alan Turing proposed that a human evaluator would judge natural language conversations between a human and a machine that is designed to generate human-like responses.

A win scenario would be as follows: a program will solve the test if it can fool half the judges into thinking it is human while interacting with them in a freeform conversation for 30 minutes and interpreting audio-visual input.

I reached out to physicist Luis Flores Castillo, member of the team that discovered the Higgs boson at CERN. He points out that "the Turing test has an interesting problem associated with it: it inherently aims at human level intelligence, and has no provision, nor expectation, for the possibility of superintelligence."

Coffee Test

AI expert, Ben Goertzel suggests a more complex test, the "coffee test", as a potential operational definition for AGI: go into an average American house and figure out how to make coffee, including identifying the coffee machine, figuring out what the buttons do, finding the coffee in the cabinet, etc.

The robot college student test

Goertzel suggests an even more challenging operational definition. That a robot should be able to enroll in a human university and take classes in the same way as humans, and get its degree.

The employment test

Nils Nilsson, one of the founders of AI, once suggested a highly demanding operational definition for AGI, the employment test. He argues that machines exhibiting true human-level intelligence should be able to do many of the things humans are able to do. Among these activities are the tasks or "jobs" at which people are employed. To pass the employment test, AI programs must... [have] at least the potential [to completely automate] economically important jobs.

This leads us squarely into one of the areas of focused research in AI: machine learning, or the ability of computers to learn by themselves.

Overall, AGI poses a deeper philosophical issue, as highlighted by Flores Castillo: "consciousness seems to me to be intimately interwoven with freedom and emotion". The question then is if AGI could truly achieve emotion, freedom, and consciousness. Or are these qualities reserved exclusively for human beings?

Machine learning

Machine learning is a type of artificial intelligence (AI) that provides computers with the ability to learn without being explicitly programmed. Machine learning focuses on the development of computer programs that can teach themselves to grow and change when exposed to new data. It is a subfield of computer science that evolved from the study of pattern recognition and computational learning theory in artificial intelligence.

Machine learning explores the construction and study of algorithms that can learn from and make predictions on data. Such algorithms operate by building a model from example inputs in order to make data-driven predictions or decisions, rather than following strictly static program instructions.

McKinsey & Co. points out that the unmanageable volume and complexity of the big data that the world is now swimming in have increased the potential of machine learning—and the need for it.

Machine learning is at the heart of automation as it enables computers and robots to perform increasingly complex tasks that were previously exclusive to humans.

Artificial Superintelligence (ASI)

Nick Bostrom, Oxford University philosopher and AI leading thinker, defines superintelligence as "an intellect that is much smarter than the best human brains in practically every field, including scientific creativity, general wisdom and social skills."

Artificial Superintelligence (ASI) can range from a computer that's just slightly smarter than a human to one that is billions or trillions of times more intelligent.

Potential existential threat

This is the existential threat that Stephen Hawking, Bill Gates, and Elon Musk among others have been talking about. It is a real threat that must be carefully addressed. Elon Musk has recently donated $10 million to the Future of Life Institute which "will look into how to keep AI friendly". Can humanity curb the potential capacity of ASI to wipe out human life?

Connecting Moore's Law with AI

What is the connection between Moore's Law and AI? Columnist Kevin Drum presents it in a sobering way, he asks the reader to consider that Lake Michigan's volume (in fluid ounces) is about the same as our brain's capacity (in calculations per second). Imagine that somehow Lake Michigan was emptied and you were given the task of filling it back to capacity (why me you might ask?). The rule is that you start by dropping a single ounce of water (about two tablespoons) into it and waiting 18 months to drop twice as much liquid. You are to keep going, doubling the amount every 18 months, until you finish the task. How long would this take?

It turns out that if you started in 1940, by year 2000 you would barely see a slight sheen on the lakebed. By 2010, only a few inches of water would have accumulated. At that point you might be tempted to think that, after 70 years of hard work, the whole effort is futile. But the shocking part is that within 15 years, by 2025, you would finish—you would fill the lake to capacity. It is mind boggling, yet math does not lie, this is the real power of exponential growth and it goes against our linear intuition.

Despite seeing the evidence of Moore's Law right in front of us, it is easy to dismiss the true power of exponential growth. That is because our intuition is linear. Right now for the most part we think that our technology is cute. Yes we can play some cool video games, we have these things attached to our hands called smartphones, and we can do a number of things with powerful desktop applications. But we have literally just seen the tip of the iceberg.

Not only useful, but incredibly powerful

Going back to the very literal metaphor of Lake Michigan, once computing power nears human-level computational power; our handheld assistants might go beyond seeming cute. We might pause to think that these things are getting incredibly smart and capable. Right now we marvel because we can identify facial features, recognize our voices, and so forth. Once our devices are capable of performing at the level of human intelligence, we will quickly realize the potential which we could only dream of before. They will drive our cars not as well as humans did, but much better. They will perform more complex knowledge-based tasks, including writing books (they already write sports columns, today). These machines will be our doctors, much more capable at diagnosing as they learn in real-time from thousands of patients treated.

These non-human abilities will mean that they are more capable than us at many functions which were reserved for humans.

Just as humanity relied on brute animal (and human) muscle power until the invention of the steam engine, we will see that the advent of computing power at the level we are about to experience, will mean that our own human brains are replaceable for many functions which were until now reserved for human beings.

From Siri to Iron Man's AI assistant

According to Peter Diamandis, in a decade, next generation AI assistants will be more like JARVIS from Iron Man, with expanded capabilities to understand and answer, thanks to work with IBM Watson, DeepMind and Vicarious, among others. "In a decade", Diamandis points out, "it will be normal for you to give your AI access to listen to all of your conversations, read your emails and scan your biometric data because the upside and convenience will be so immense."

Quantum computing

A quantum computer harnesses phenomena explained by quantum physics, such as superposition and entanglement, to perform operations on data, which are impossible with the traditional binary approach. "With quantum computing, a quantum bit or qubit exists in both of its possible states at once, a condition known as a superposition", as explained by AI expert Ciara Byrne. "An operation on a qubit thus exploits its quantum weirdness by allowing many computations to be performed in parallel. A two-qubit system would perform the operation on 4 values, a three-qubit system on 8 and so forth."

Andris Ambainis, computer scientist active in the fields of quantum information theory and quantum computing, gives an example: "Let's say that we have a large phone book, ordered alphabetically by individual names (and not by phone numbers). If we wanted to find the person who has the phone number 6097348000, we would have to go through the whole phone book and look at every entry. For a phone book with one million phone numbers, it could take one million steps. In 1996, Lov Grover from Bell Labs discovered that a quantum computer would be able to do the same task with one thousand steps instead of one million."

The next step for Moore's Law?

Following Moore's Law sometime within the next two decades we will find that we can no longer shrink the circuits on a microprocessor already measured on an atomic scale, currently at 10 nanometers. Hence the next step would be quantum computing, which would perform calculations significantly faster than any silicon-based computer.

According to Jerry Chow, manager of IBM's Experimental Quantum Computing group, quantum computing has potential for things like drug discovery, drug design (testing drug combinations by the billions at a time), chemical design, and hopefully applications in the bio-pharma realm. Other possible applications include solving complex physics problems which are currently beyond our understanding and creating unbreakable encryption through the use of quantum cryptography.

We are making steady progress towards the availability of quantum computing, regarded as the "holy grail" at the intersection of physics, quantum mechanics, and computing. When could we expect to see a functional quantum computer? Robert Schoelkopf, Researcher at the Department of Applied Physics at Yale University states that "it is hard to estimate how long it will be until we have functional quantum computers, but it will be sooner than we think." When quantum computers go mainstream we will enter a new era of computing that was, not long ago, only science fiction.

3

Autonomous Vehicles

THE NOT-SO-DISTANT FUTURE OF
TRANSPORTATION

*All of the biggest technological inventions created by man —
the airplane, the automobile, the computer — says little about
his intelligence, but speaks volumes about his laziness.*
—*Mark Kennedy (Author)*

On June 16, 1897 in Stuttgart, the world's first gasoline-powered taxicab, the Daimler Victoria, was delivered to German entrepreneur Friedrich Greiner. The eight-horsepower engine afforded a top speed of 24 kilometers per hour. The vehicle came equipped with a taximeter, a six-year old innovation which allowed fares to be calculated based on distance covered and waiting time. Within three years Greiner ordered another half dozen taxis, launching the world's first motorized taxi company. By the end of the 19th century, motorized cab companies would proliferate globally, and by 1920 they were a ubiquitous part of urban life.

Fast forward 113 years. On May 31, 2010, two young techies, Garrett Camp and Travis Kalanick launched a service in San Francisco, a smartphone app connecting a driver with someone looking for a ride. Uber was born and with it a service that has completely disrupted the taxi industry. Uber and similar services have significantly enhanced the interaction between drivers and passengers; the proof is that users are typically delighted.

Taxi industry disruption

This clear preference has taken its toll on the incumbent taxi industry, sometimes devastatingly so. Point in case is taxi tycoon Evgeny "Gene" Freidman, estimated to own some 1,000 New York City taxi medallions which he leases to drivers. At the peak of the market, around 2013, each medallion was worth an estimated $1.1 million. However, mainly thanks to Uber, his assets have depreciated around 40 percent. The New York Times reports that Freidman is currently locked in litigation with Citibank, trying to seize 87 medallions from him, following unpaid loans. Overall the problem for NYC taxi owners is not only asset depreciation but that creditors are less willing to lend to potential owners. Medallion sales have stagnated in a disrupted market; lenders are less willing to provide financing. Freidman argues that he is too big to fail and this is a public problem, demanding that the City of New York step in to guarantee a "normal" value.

The challenge really is that the playing field has been fundamentally disrupted by technology. The licenses he owns, which skyrocketed in price thanks to limited supply, are no longer as demanded as in the pre-Uber era. Aided by the advent of smartphones and app ecosystems, Uber and similar services have taken a big bite out of the Big Apple taxi market in a few short years.

Uber domination

Forbes reports that, based on expense reports filed by business travelers in the first quarter of 2015, "an average of 46 percent of all total paid car rides were through Uber in major markets across the U.S." During the same period, rides in taxis, limos and shuttles also took a hit, falling 85 percent.

More shockingly, Business Insider UK reports that in San Francisco—Uber's most mature market—the company is already three times the size of the local taxi market ($500 million vs. $140 million in yearly revenue). Uber's rides in San Francisco are growing threefold per year, in New York, four times, and in London, five to six times per year.

Today Uber is worth over $50 billion, *half* of the entire global taxi industry (estimated by Forbes at $100 billion), most strikingly, without owning a single taxi. Not bad for a five year old company.

Enabling platform

One of the events that enabled Uber's rise can be traced back to June 29, 2007. On this day Steve Jobs unveiled the first generation iPhone, describing it as a "revolutionary mobile phone". A year later, on July 10, 2008, Apple launched the App Store (iOS), opening the gates for an ecosystem of apps. That was followed by the Android ecosystem, furthering the expansion of the app world in mobile computing. We could not understand Uber's success without those two platforms: smartphones and the app ecosystem. We will revisit this concept in the second part of the book, where we discuss the combinational nature of exponential technologies in novel ways.

No going back for users

While not everything is blue skies for Uber due to backlash from incumbents and ensuing regulation, for users, regardless of the outcome, there is no going back. Passengers are now used to it, and clearly prefer it over the traditional alternative. The ride-sharing platform has completely disrupted an industry that was 113 years old. But there is a more profound shift literally around the corner.

Driverless vehicles

Hold on to your seat. If we are seeing protests because of Uber, wait until the taxi shows up *without* a driver. On July 8, 2015, Uber announced that it would buy half a million automated vehicles from Tesla—the company is on its way to replace the driver, Uber's human component. Major automotive manufacturers, including Audi, Mercedes-Benz, Nissan, and Honda, to name a few, along with

Google and Tesla, are working on some version of an autonomous vehicle, that is, a self-driving car.

What are autonomous vehicles?

An autonomous vehicle also known as a driverless car, self-driving car, or robotic car is an automated or autonomous vehicle capable of fulfilling the main transportation capabilities of a traditional car. As an autonomous vehicle, it is capable of sensing its environment and navigating without human input.

Autonomous vehicles sense their surroundings with such techniques as radar, lidar (light detection and ranging), GPS, and computer vision. Advanced control systems interpret sensory information to identify appropriate navigation paths, as well as obstacles and relevant signage.

By definition, autonomous vehicles are capable of updating their maps based on sensory input, allowing the vehicles to keep track of their position even when conditions change or when they enter uncharted environments.

Driverless taxis

Forbes reports that about three quarters of the fare charged by Uber goes to drivers' wages, i.e. the human component. What happens if and when the driver is no longer needed? Then all of a sudden, the taxi becomes highly affordable to everyone.

Futurist Zack Kanter cites a Columbia University study which suggests that with a fleet of only 9,000 autonomous cars, Uber could replace every taxi cab in New York City. Passengers would wait an average of *36 seconds* for a ride costing around $0.50 per mile.

Even today, for some urban dwellers taking a taxi is cheaper than the total cost of owning a car. When taxis go driverless, people will have less incentive to own a car.

At present, cars are one of the most under-utilized assets; they sit idly, parked, about 96 percent of the time, according to Morgan Stanley. We use our cars 4 percent of their useful life! If you add licensing costs, fuel, maintenance, cost of money, and other expenses, it is evident that they are expensive assets, and that the majority of users would be delighted to do without them.

However, the impact of driverless cars extends way beyond the taxi industry.

Driverless future

When an individualized taxi service becomes as cheap as taking the bus, city dwellers will buy fewer and fewer cars. A study by the University of California at Berkeley concluded that vehicle ownership among car sharing users was cut in half.

You will hail a cab with your smartphone app with the only difference that it will be self-driven and it will be much cheaper. In essence we will go from an ownership model to a rental (pay-as-you-go) or even subscription-based transportation model. This will have huge implications.

Reduced vehicle fleet

First of all this eliminates the inefficiencies caused by a bloated fleet of private vehicles; a reduced number of cars can service a given urban area, because a self-driving car doesn't need to be parked. It can go pick up the next passenger as soon as it drops you at your destination. Taxi-like driverless services are destined to become a primary mode of transportation in urban settings. Morgan Stanley estimates that a 90 percent reduction in crashes would save nearly 30,000 lives and prevent 2.12 million injuries annually.

Direct impact

Six million professional drivers are employed driving a truck, bus, or a delivery vehicle in the U.S. alone. Car manufacturing employs another 900,000 or so. Driving is the most popular job in the majority of American states. Widespread adoption of driverless vehicles could decimate most of these jobs by 2035.

Demise of ancillary industries

As the driverless trend accelerates, other structural changes will follow. A smaller state apparatus will be needed for issuing driver licenses, there will be a diminishing need for traffic cops (unless "robot" drivers start to misbehave). Traffic wards, parking attendants, and other jobs associated with human driving will be progressively phased out of existence. The parking industry is worth $100 billion in annual revenue. However that is just the tip of the iceberg.

Kanter points out that the automobile insurance market is worth $198 billion, the automotive finance market, $98 billion, and the automotive aftermarket, $300 billion. Demand for these would plummet.

In the trucking industry there are a number of ancillary services including service stations, motels, catering to the needs of truck drivers. Demand for these could drop sharply.

The transition

Governments will undoubtedly face increasing pressure to slow down or outright ban autonomous driving vehicles. But sooner or later, as the advantages prove to outweigh the costs, society will transition to this automated model. However, make no mistake, this will be a painful transition, and steps must be taken to soften the fallout.

4
Flying drones
OPENING THE SKIES

A few months ago I was hiking in the beautiful west of Ireland in Shramore, near Newport, Co. Mayo with Frank McManamon. During the trek, I was astounded when he demonstrated how Sam, his sheepdog, obeyed six or so different whistle commands. I half-jokingly said to him that in the not-so-distant future Sam could be out of work, replaced by a drone which would let Frank do his sheep-herding from a distance, maybe even from home. It turns out it doesn't have to be a (lame) joke.

In Dannevirke, New Zealand, Michael Thomson, a young engineering student, has been busy herding sheep on his sister's farm with a "quadcopter" he crafted himself. The drone turned out to be more effective at sheep-herding than anyone anticipated, "much quicker than horses and dogs," Thomson explained to the Wall Street Journal and local media outlets. "I didn't have to worry about fences or going through gates, I'd just fly over them," he described.

From lifeguard drones, to aerial photography, and flying medical supplies in remote areas, drones have become increasingly useful gadgets.

What are drones?

An unmanned aerial vehicle (UAV), commonly known as a drone, and also referred to as an unpiloted aerial vehicle and a remotely piloted aircraft (RPA) by the International Civil Aviation Organization (ICAO), is an aircraft without a human pilot aboard. Drones range in size from a few centimeters to tens of meters in length. Drones are either controlled by a person on the ground or autonomously via a computer program. First popularized due to their military use, drones are now used for everything from wildlife and atmospheric research to disaster relief and photography.

Drone era

Drones have ushered in an era of not only hobbyists but of professionals using them to accomplish increasingly significant tasks.

Lifeguard drone

The Pars Aerial Rescue Robot is an aerial drone for sea rescue missions. RTS, the company behind it, states that the technology decreases the time needed to reach drowning people and that could also save more than one sinking person at a time. Lifeguards or rescue team members can fly the drone over the victims with a radio control. The drone, carrying cameras and rubber life tubes, can be sent out by lifeguards from the shore, helping rescue drowning individuals quicker than with human assistance only.

Drone photography

Professional photographers find drones extremely useful for shots that were previously impossible. From creative snapshots of weddings to astounding photographs of previously inaccessible scenery, drone photography has become the new normal. You would expect a social network for flying drone photographers, right? Enter Dronestagram, the first social network for this market niche, giving professional drone photographers and enthusiasts a place to upload breathtaking aerial photographs that can be captured using their UAV's. Dronestagram recently partnered with National Geographic, GoPro, Adobe and other notable companies for the 2015 Drone Aerial Photography Contest with very aesthetically pleasing results.

Another company, Falkor Systems, has targeted extreme sports enthusiasts, focusing on skiers and base-jumpers. "The angles people get [while filming] are not quite as intimate as would be possible with an autonomous flying robot," commented Falkor CEO, Sameer Parekh. He envisions a small UAV device that can accompany a downhill skier.

Air parcels

What if you could order something online and get your hands on it in a matter of *minutes*? Amazon is testing a 30-minute delivery service with Amazon Prime Air. The order would have to be less than five pounds (2.26 kg), which, according to Jeff Bezos,

Amazon CEO, includes 86 percent of the packages Amazon currently sells. Earlier this year the FAA issued the company a special "experimental airworthiness certificate", allowing Amazon to conduct outdoor research, testing, and training.

Flying medical supplies and disaster relief

Flirtey is the Australian drone start-up that became the first company to receive approval from the FAA to deliver packages by unmanned aerial vehicle (UAV) in the U.S.

The company will deliver medical supplies to a rural coal mining region in Wise County, Southwest Virginia, not only one of the most isolated places in the U.S. but on the planet. For some 3,000 residents the isolation means precarious access to healthcare.

Flirtey will tether medical supplies to six-rotor drones, delivering 24 packages of prescription medication, weighing some 10 pounds (4.5kg) each to Wise County.

"This is the first step in proving that on-demand drone delivery can revolutionize the way medical care can be delivered to remote communities, and eventually from your local pharmacy to your front door," said Flirtey Co-Founder Tom Bass. "This will be a game changer for millions in America."

In a similar effort, a team from the Harvard-MIT Division of Health Sciences and Technology received a grant from the Bill and Melinda Gates Foundation to develop drones that deliver vaccines and medicines to remote locations and disaster zones. According to a Forbes report, "They envision healthcare workers using the drones via mobile phone to swiftly get medicine to people in isolated areas in a cost-effective way, helping to improve the coverage of vaccine supplies."

Highway monitoring

In the state of Georgia, U.S. a project is underway using drones to inspect roads and bridges, surveying land with laser mapping and alerting authorities to traffic jams and accidents.

Javier Irizarry, director of the CONECTech Lab at the Georgia Institute of Technology stated that "drones could keep workers safer because they won't be going into traffic or hanging off a bridge."

Challenges

While the potential use for drones is limited by imagination, not all uses of technology are as harmless. A recent video that went viral on the Internet showed a handgun attached to a hovering drone being remotely fired. This clearly raises red flags that must be addressed.

Moreover flying drones have become an issue of airspace security as they can interfere with passenger aircraft. While drones are highly useful, regulation that ensures safety for everyone is a must. The trick is to achieve this while not hindering industry growth, limiting the potential benefits of this technology.

5
3D Printing
MANUFACTURING REVOLUTION

On the morning of November 25, 2014 orbiting some 400 kilometers above Earth on board the International Space Station (ISS), NASA astronaut Barry "Butch" Wilmore checked the output of his printer. A day earlier, Houston ground control had sent it a command to print a plate which read: "MADE IN SPACE NASA". Astronaut Wilmore inspected it and was satisfied that they had just 3D-printed humanity's first object in space. If you consider the amazingly complex and expensive ISS resupply process, one can appreciate the fact that NASA found a way to wirelessly "beam" an object onto the ISS, saving hundreds of millions of dollars per event. On that date we entered an era allowing us to conceive objects on Earth and replicate them in three-dimensional form outside our planet.

Tipping point

In a Harvard Business Review article, Richard D'Aveni states that 3D printing is at a tipping point, fast approaching the 20 percent adoption rate which

Gartner analysts consider as the benchmark for mainstream adoption.

What is 3D printing?

A 3D printer works much like a conventional inkjet printer, except that the output is not a flat, or two-dimensional printout, but a three-dimensional object. The technology typically works by depositing successive layers of a material such as plastic resin in intricate patterns. The resin solidifies enabling the production of a wide array of objects, including those that are difficult to produce via conventional manufacturing processes. 3D printing is also known as additive manufacturing because an object is made by adding successive layers as opposed to cutting away or shaving off excess material. It is not a new technique, as it has been used for over 25 years to create industrial prototypes, however, it is about to go into the mainstream consumer market.

From rapid prototyping to mass production

3D printing also known as additive manufacturing has been around since the 1980s. However, it was not capable enough or sufficiently cost-effective for the majority of end-product or high-volume commercial applications. But that is quickly changing.

Typically there has been two price points for 3D printers: costly high end 3D printers and inexpensive yet less capable printers. However, according to a

PricewaterhouseCoopers report a new category is emerging: a mid-level 3D printer which offers many high-end features in a desktop form factor while being more economical. The added capabilities at a more reasonable price will make the technology more useful and speed its adoption.

Today most 3D printers utilize only one type of material at a time. This can be plastic, metals, including gold, silver, bronze, titanium, and ceramics, and more recently wood. According to PricewaterhouseCoopers, 3D printers must be improved in three areas to capitalize beyond rapid prototyping. The first one is performance, including speed, resolution, and autonomous operation, ease of use, reliability, and repeatability. Secondly, multi-material capability and diversity, including the ability to mix materials while printing a single object. Lastly increasing the variety of finished products, such as the ability to 3D print embedded sensors, batteries, and electronics.

The key insight is that 3D printers will be able to produce increasingly complex objects, at greater speeds, and with greater functionality.

Whereas today 3D printing may be perceived by the public as being limited to producing arts, crafts, and spare parts, with these key developments we will have complex products printed increasingly faster. Given enough time we will be able to print anything we can imagine.

Canalys, a market research firm, predicts that the global market for 3D printers and services will grow from $2.5 billion in 2013 to $16.2 billion in 2018, an astounding year on year growth of 45.7 percent.

Supply chain disruption

What if we can completely bypass the complexities of supply chain logistics? What if increasingly localized 3D printing machines could fulfill the needs of smaller cities or villages? Imagine that, instead of paying 30 percent or 40 percent of the cost of a product on freight, you could 3D print a wide variety of objects at home or at specialized kiosks within 25 kilometers of smaller consumer centers. This could spell disruption for global supply chain management. Rick Smith, 3D printing entrepreneur and Forbes contributor, references this as direct digital manufacturing, the capacity to manufacture components and finished goods near their point of use.

Another advantage over conventional mass production is the fact that you can custom produce your products, tailored to suit individual consumers. This creates an added value in terms of personalization. You don't need to have the same mass product as everyone. You can create truly unique products with your name on them, at no additional cost, opening the door for a new type of affordable customized production.

Gartner predicts that 3D printers with the anticipated potential described above will be available for less than $1,000 by 2016. Printer improvements will likely accelerate in the next few years. As with every technology it will also become increasingly more automated and easier to use.

Current and potential applications

PwC identifies some further potential uses of 3D printing, including the following: in automotive and industrial manufacturing, to consolidate many components into a single complex part, to create production tooling, produce spare parts and components, to create complex geometry parts not possible with traditional manufacturing.

In health care, to plan surgery using precise anatomical models based on CT scan or MRI, to develop custom orthopedic implants and prosthetics. Some makers of hearing aids and dental braces have adopted the technology for finished products.

In retail, to create custom toys, jewelry, games, home decorations, and other products, to print spare or replacement parts for auto or home repair, for example.

In sports, to create custom protective gear for better fit and safety, to create custom spike plates for soccer shoes based on biomechanical data.

3D printed house

Researchers in Sweden are experimenting with 3D printed houses. In a collaborative EU-funded endeavor, the project aims to 3D-print houses with cellulose. "The idea of the project is to develop a technology that can be used in reinforcing the manufacturing industry in the region," explains Marlene Johansson, director of Sliperiet.

The +Project looks to target industries within the wood and construction sectors, as well as within design, architecture and even IT. It has goals of producing cellulose-based 3D printable materials which can be used to print everything from walls within a home, to simple weather-stripping, doors and even entire houses.

"There is already technology in place to print parts of houses in concrete, for instance," explains Linnéa Therese Dimitriou, Creative Director at Sliperiet. "Now, with this project, the region is one step closer to the front edge in the area of digital manufacturing and so-called mass-customization. This opens up for incredibly exciting future opportunities for the regional forest and construction industry as well as for regional raw material."

Bridge printed "on the go"

In Amsterdam, 3D printing technology will enable a pedestrian bridge to be 3D printed. A Dutch startup company, MX3D, is planning to be the first to print a steel bridge across a canal in Amsterdam. According to the company, the robotic arms, to be placed on opposite ends of the bridge, will heat metal to 2,700 degrees Fahrenheit and weld together metal pieces one by one. Moreover, the robotic arms use the bridge's own structure as support, eliminating the need for additional work space.

According to Tim Geurtjens, MX3D Chief Technology Officer, "what distinguishes our technology from traditional 3D printing methods is that we work according to the 'Printing Outside the box' principle. By printing with 6-axis industrial robots, we are no longer limited to a square box in which everything happens. Printing a functional, life-size bridge is of course the ideal way to showcase the endless possibilities of this technique."

3D printing on course to deliver

According to PricewaterhouseCoopers, the current pace of innovation suggests that the high expectations set for 3D printing will be met. The 3D printer market is transforming rapidly. Robust innovation at established vendors and among entrepreneurs and hobbyists is providing a test ground for filling the market with more midrange systems that bring enterprise-class capabilities at much lower prices.

6

Internet of Things
EVERYTHING CONNECTED

The real danger is not that computers will begin to think like men, but that men will begin to think like computers.
— Sydney Harris (Journalist)

What will happen when all your things start *talking* amongst each other? According to Gartner some five billion things are connected to the Internet, making up the Internet of Things (IoT), up 30 percent from 2014, reaching 25 billion within five years.

This sudden expansion will boost the economic impact of the IoT as consumers, businesses, government authorities, hospitals and many other entities find new ways in which to make use of the technology.

The Internet of Things (IoT) has become a powerful force for business transformation, and its disruptive impact will be felt across all industries and all areas of society.

What is the Internet of Things?

The Internet of Things (IoT) is the network of physical objects that contain embedded technology to communicate and sense or interact with their internal states or the external environment. This allows everyday objects have network connectivity, allowing them to send and receive data

Consumer applications will drive the number of connected things, while enterprise will account for most of the revenue. Gartner estimates that 2.9 billion connected things will be in use in the consumer sector in 2015 reaching over 13 billion in 2020. The automotive sector will show the highest growth rate at 96 percent in 2015.

A trillion-sensor economy

Peter Diamandis estimates that by 2025, the IoT will exceed 100 billion connected devices, each with a dozen or more sensors collecting data. This will lead to a trillion-sensor economy driving a data revolution beyond our imagination. A report by Cisco estimates the Internet of Things (also known as Internet of Everything) will generate $19 trillion of newly created value.

A sensor for everything

Consider that under the view of IoT architects, objects, animals, and people can and will be provided with unique identifiers, giving both living and nonliving entities the ability to transfer data over a network without requiring human-to-human or human-to-computer interaction. The implications could be groundbreaking.

Sample benefits in a manufacturing context

According to Manufacturing Global, increased product knowledge is one of the benefits that IoT can deliver. "By connecting a product to the Internet, a manufacturer should expect to receive granular, real-time information about how it is being used." Another key area is the improvement of customer service. As the manufacturer has a greater ability to identify problems, customers can expect a greater level of customer service, providing hints and tips on how to get the most out of a product. This can also represent an opportunity for greater interactivity with the consumer.

Entrepreneur and Huffington Post contributor, Paul Mashegoane, sees the Internet of Things as a game changer and argues that IoT possibilities are "endless; your smartphone may know about your health more than your doctor does, your fridge may know more about your diet than you know. A traffic system will get to a point where it will control itself to better monitor traffic, send the information to your car to navigate the city more easily; collect the data of what's happening on the road of accidents and traffic congestions, then send it over to the transport department."

Exciting uses

Futurist and tech consultant Daniel Burrus paints a vivid picture of potential IoT uses: "In 2007, a bridge collapsed in Minnesota, killing many people, because of steel plates that were inadequate to handle the bridge's load. When we rebuild bridges, we can use smart cement: cement equipped with sensors to monitor stresses, cracks, and warpages. This is cement that alerts us to fix problems before they cause a catastrophe. And these technologies aren't limited to the bridge's structure."

Burris explains that if there is ice on the bridge, the same sensors in the concrete could detect it and communicate the information via the wireless Internet to your car. Once your vehicle knows there's a hazard ahead, you can slow down or avoid the road altogether, turning data into actionable insights.

Furthermore smart cars can connect to smart city grids and start talking to each other, optimizing traffic flow. "Instead of just having stoplights on fixed timers, we'll have smart stoplights that can respond to changes in traffic flow. Traffic and street conditions will be communicated to drivers, rerouting them around areas that are congested, snowed-in, or tied up in construction," explains Burrus.

Privacy issues

However, the IoT could also result in huge privacy headaches that we must be aware of. "If privacy isn't dead yet, then billions-upon-billions of chips, sensors, and wearables will seal the deal," say Jat Singh and Julia Powles, researchers at the University of Cambridge. Yet there are ways around this. "If there is the vision and commitment to realizing pervasive computing in a way that is open, diverse, innovative, and high-value, then privacy may just stand a fighting chance." This is an issue that must be thoroughly reviewed and addressed to prevent security breakdowns.

Key areas of development

To fully advance the potential of the Internet of Things, McKinsey & Co. identifies five key areas of development. These are: interoperability of the different types of sensors and software; the need for low-cost, low-power hardware; ubiquitous connectivity—that is the need for Internet everywhere; analytics software *and analytic people* needed to help make sense of it all; and more attention needs to be paid to privacy, confidentiality, and security, as pointed out previously.

7

Virtual and Augmented Reality
THE GIFT OF TELEPRESENCE

*The art challenges the technology, and the technology
inspires the art.*
—John Lasseter (Director)

It's 8:59 am. You are about to kick off your weekly staff meeting. You look around: everyone is ready to go. The usual is discussed: project status, KPI's, holiday plans, and the latest episode of *Game of Thrones.* Forty five minutes into the meeting you are done: all fourteen team members agree on key objectives and results—you are all set for the week. You go back to your desk and look out the window. You are relaxed by the live waterfall set against a jaw-dropping view of the forest. It doesn't get better than this.

Rewind 60 minutes: you wake up at 8:30 at your lakeside cabin, 100 miles from the nearest city and 2,500 miles from company headquarters. You stay in bed for 15 minutes. You get up and make a pot of coffee slowly enjoying every sip. At 8:57 you sit at your home office, strap on your virtual reality headset, and log in. The meeting starts promptly at 9am.

Thanks to the latest advances in VR you felt as if you were actually at the company meeting. The funny thing is that half of your staff was also working from home. It will become increasingly hard to tell the difference between a *real* meeting and a virtual one.

What is Virtual Reality?

Virtual Reality is a computer-generated simulation of a three-dimensional image or environment, in which interaction takes place using special electronic equipment, such as a helmet, or goggles, or other eye wear, and may include other sensors. The environment can be explored and interacted with by a person. That person becomes part of this virtual world or is immersed within this environment and whilst there, is able to manipulate objects or perform a series of actions. VR is presented to the user in such a way that the user suspends belief and accepts it as a real environment. On a computer, virtual reality is primarily experienced through two of the five senses: sight and sound.

The big new platform

In a public announcement in June 2015, Mark Zuckerberg expressed the view that he believes that immersive experiences like VR will become the norm. This is to be expected as Facebook acquired leading VR company Oculus Rift. Taking

Zuckerberg's increasing levels of excitement around VR, we can trust that he has virtually seen the future, if you pardon the pun.

Furthermore, he envisions that we will use augmented reality technologies including devices that we can wear almost all the time to improve our experience and communication.

When you put on a VR headset such as Oculus Rift's, you enter a completely immersive computer-generated environment, like a game or a movie scene or a place far away. The incredible thing about the technology is that you feel like you're actually present in another place with other people. People who try it say it's different from anything they've ever experienced in their lives.

Oculus's mission is to enable you to experience the impossible. Their technology opens up the possibility of completely new kinds of experiences.

Zuckerberg explains that immersive gaming will be the first domain they will target, as the Rift is already highly anticipated by the gaming community.

VR will become a platform for a multitude of experiences. "Imagine enjoying a court side seat at a game, studying in a classroom of students and teachers all over the world or consulting with a doctor face-to-face—just by putting on goggles in your home," remarks the Facebook founder.

The possibilities are seemingly endless. From gaming, to next generation social networking, to entertainment and business applications, VR has the potential to usher in transformative change.

Indistinguishable from reality

In the VR front we see technology getting so good that it will become indistinguishable from reality. This has many powerful implications.

Peter Diamandis points out game changers like Magic Leap (in which Google just invested over $500 million) that are developing technology to "generate images indistinguishable from real objects and then being able to place those images seamlessly into the real world."

Nine new VR experiences will be premiering at the Sundance Film Festival this year, spanning from artistic, powerful journalistic experiences like Project Syria to full "flying" simulations where you get to "feel" what it would be like for a human to fly.

This is truly a new communication platform, Zuckerberg says. "By feeling truly present, you can share unbounded spaces and experiences with the people in your life. Imagine sharing not just moments with your friends online, but entire experiences and adventures". Facebook is committed to making this kind of immersive, experience a part of daily life for billions of people.

Augmented Reality

Going in a direction similar to that of virtual reality is augmented reality.

What is Augmented Reality?

Augmented reality (AR) is a technology that superimposes a computer-generated image on a user's view of the real world, thus providing a composite view. For example, super imposed on a helmet, windscreen, or regular eye wear a user could see text alerts, or read a map directions, without looking at a separate screen. By contrast, virtual reality replaces the real world with a simulated one. With the help of advanced AR technology the information about the surrounding real world of the user becomes interactive. Artificial information about the environment and its objects can be overlaid on the real world.

Life seen through different lenses

Whereas virtual reality immerses you in a fully computer-generated environment, augmented reality (AR) overlays text and graphic information over a display or even holographically.

Microsoft HoloLens for example is the "first fully untethered, see-through holographic computer," according to Microsoft. "It enables high-definition holograms to come to life in your world, seamlessly integrating with your physical places, spaces, and things." They call this experience mixed reality. Holograms mixed with your real world will unlock all-new ways to create, communicate, work, and play.

As of 2015, a number of Windows Holographic applications have been showcased or announced, for the HoloLens. These include HoloStudio, a 3D modelling application which can produce output for 3D printers; an implementation of the Skype telecommunications application; an interactive digital human anatomy curriculum by Case Western Reserve University and Cleveland Clinic, architectural engineering software tools by Trimble Navigation, a version of the video game Minecraft, and OnSight and Sidekick, software projects being developed as part of a collaboration between NASA and Microsoft to explore mixed reality applications in space exploration.

Augmented and virtual reality

Diamandis points out that billions of dollars invested by Facebook (Oculus), Google (Magic Leap), Microsoft (Hololens), Sony, Qualcomm, HTC and others will lead to a new generation of displays and user interfaces.

The screen as we know it, on your phone, your computer and your television, will disappear and be replaced by eyewear. The result will be a massive disruption in a number of industries ranging from consumer retail, to real estate, education, travel, entertainment, and the fundamental ways we operate as humans. We will review this more in the next part.

8

Biotechnology
EDITING NATURE

This is the whole point of technology. It creates an appetite
for immortality on the one hand. It threatens universal
extinction on the other. Technology is lust removed from
nature.
—Don DeLillo (Author)

English novelist Daniel Defoe, author of *The Political History of the Devil* (1726) is credited as the first one to quote the certainty of "death and taxes". But what if death was not as unquestionable as we all think it is? Could there be someone out there trying to "cure" death? As crazy as it seems, it turns out this is not the quest of a mad scientist, but a company backed by Google called Calico.

Calico: a project to "cure" death

Calico, the California Life Company, is Google's moonshot project in healthcare and biotech aimed at reversing the processes by which the body ages. While the company is still far from offering a market-ready solution, there has been some progress made in relation to telomeres, tiny caps at the end of each strand of our DNA. It turns out that these may hold the key to discovering the proverbial fountain of youth. Scientists have discovered that each time our cells reproduce, telomeres get progressively shorter, which is, in short, the cause of age-related cellular breakdown. In 2009, a group of scientists was awarded the Nobel Prize for their discovery of how the enzyme telomerase impacts telomere length. This is an area into which companies like Calico are researching, aiming to reverse aging. While aging may seem a moonshot for Google, it also has become a matter of computing power and machine learning. "Digital insights are becoming increasingly important in the life sciences, because genetics is a data-driven practice, giving us a real place to add value," says Dr. Krishna Yeshwant, a general partner with Google Ventures.

What is Biotechnology?

Biotechnology is the use of living systems and organisms to develop or make products, or "any technological application that uses biological systems, living organisms or derivatives thereof, to make or modify products or processes for specific use" (UN Convention on Biological Diversity, Art. 2). Over the past 30 years, biologists have increasingly applied the methods of physics, chemistry and mathematics in order to gain precise knowledge, at the molecular level, of how living cells produce substances such as medicines, food, and fuel.

Cut and paste DNA editing

In a similar front in the biotech world, scientists have found a way to easily and affordably cut and paste DNA segments in a process similar to cutting and pasting text. CRISPR-Cas9 is a technique invented by Jennifer Doudna (director of UC Berkeley's Innovative Genomics Initiative) and Martin Jinek in 2012 at UC Berkeley which is at the center of genomics. It is an acronym for "clustered regularly interspaced short palindromic repeats," a description of the genetic basis of the method; Cas9 is the name of a protein that makes it work. This technique makes it easy, affordable, and quick to move genes around, in any living organism, from bacteria to people. CRISPR-Cas9 opens exciting possibilities in the field of genome engineering. However, Doudna, and other genomics notables, "strongly discourage any attempts at making changes to the human genome that could be passed on to offspring". We have entered the era where it is not only possible to sequence the whole human genome to understand the code we are made of, but we are now capable of *re-writing* our own code, our DNA, for a number of purposes. We are running into unchartered territory in the world of biotech that may unlock all sorts of promises, some of which may be highly beneficial to humanity and others which we may be completely unaware of.

Printing live human tissue

In another front, Organovo has taken 3D printing to a whole new level, by becoming a leading innovator in bioprinting, that is, using *live* cells as ink to create three-dimensional living designs. This is functional human tissue, proven to function like native tissues. The printed tissue accurately represents human biology, enabling ground-breaking therapies and applications. Keith Murphy, CEO of Organavo, explains that the company has several research partnerships with firms like L'Oreal and Merck & Co., to help them test their cosmetics on 3D printed tissue as opposed to testing with animals.

Organovo, for example, has also succeeded in creating 3D printed human kidney tissue, on which they tested a new drug called Rezulin. The results demonstrated within seven days that the drug caused significant damage to the cells, dramatically speeding up the testing process. This is a breakthrough allowing for quicker routes for drug development.

Bioprinting methods

According to a report by PwC, bio-printing typically uses two inks. One is the biological material and the other is hydrogel that provides the environment where the tissue and cells grow. The breakthrough to add blood vessels was the development of a third ink that has an unusual property: it melts as it cools, not as it warms. This property allowed scientists to print an interconnected network of filaments and then melt them by chilling the material. The liquid is siphoned out to create a network of hollow tubes, or vessels, inside the tissue. Such creations are possible only with 3D printing, generating new possibilities beyond traditional manufacturing.

Drawing blood

Elizabeth Holmes is heralded as the youngest self-made billionaire, founder of Theranos, a biotech company that has developed novel approaches for laboratory diagnostic tests using significantly less blood. The company's blood-testing platform uses a few drops of blood obtained via a finger-stick rather than vials of blood obtained via traditional venipuncture.

Theranos' device design uses a fingerstick to draw a microliter sample of blood into a disposable cartridge, which is loaded into the device's "reader" for analysis; results are sent wirelessly from the reader to a secure database, from where they go online to the patient or patient's physician. The company's device claims include that the results will be received faster than the usual three-day delay for centralized laboratory testing. Theranos offers to perform up to 30 tests with a single sample, while eliminating human errors associated with handling and delays associated with traditional tests. Theranos is valued today at around $9 billion.

Disruption of healthcare

Peter Diamandis sees that, overall, existing healthcare institutions will likely face disruption as new business models with better and more efficient care emerge. Thousands of startups, as well as today's data giants (Google, Apple, Microsoft, SAP, IBM, etc.) will all enter this lucrative $3.8 trillion healthcare industry with new business models that dematerialize, demonetize and democratize today's inefficiencies.

Biometric sensing (wearables) and AI will make each of us the CEOs of our own health. Large-scale genomic sequencing and machine learning will allow us to understand the root cause of cancer, heart disease and neurodegenerative disease and what to do about it. Robotic surgeons can carry out an autonomous surgical procedure perfectly (every time) for pennies on the dollar. Each of us will be able to regrow a heart, liver, lung or kidney when we need it, instead of waiting for the donor to die.

9
Alternative Energies
POWERING OUR WORLD

On May 1, 2015, billionaire entrepreneur Elon Musk, announced a "fundamental transformation" in the energy landscape. "We have this handy fusion reactor in the sky called the sun, you don't have to do anything, it just works, shows up every day, and produces ridiculous amounts of power." He continued: "The problem with existing batteries is that they suck", pointing out that they are expensive, unreliable, they have poor integration, poor lifetime, they are not scalable, and they're unattractive. Most importantly, he noted society's reliance on contaminating fossil fuels.

Tesla's Powerwall

The Powerwall by Tesla is an answer to solving the energy problem. It is, in essence, an energy storage device—a lithium ion battery borrowed from Tesla's car battery technology—that captures sunlight and stores it for residential consumption. It is "completely automated, it installs easily and requires no maintenance," according to Tesla. During sunlight hours, the home battery stores surplus electricity generated from solar panels installed at the home or from the utility grid when rates are low. It converts direct current electricity into the alternating current used by residential lights, appliances, and devices.

The Harvard Business Review points out, though, that Tesla's isn't the first battery for energy storage on the market, but it may be the first that is simple to use, install, and maintain. The other aspect that is noted is the creation of an energy ecosystem: solar panels, home batteries, and electric vehicles can interact in a compatible plug-and-play ecosystem, benefitting from economies of scale, and possibly contributing to a consumer trend.

The Washington Post notes that "battery-storage technologies will improve so much over the next two decades that homes won't be dependent on the utility companies".

Cheap, abundant energy?

Does the idea of cheap and abundant energy sound like a pipe dream? It may turn out that it is not utopia. A report by Citi titled "Evolving Economics of Power and Alternative Energy" published in March 2014 states that "Renewable Energy is Increasingly Cost Competitive: Renewables energy, primarily solar and wind, costs continue to decline and are increasingly competitive with natural gas. The drivers of the lower costs are lower construction costs, higher efficiencies, and favorable financing terms".

The quest for alternative energies is not limited to solar. Wind, biomass, thermal, tidal, and waste-breakdown energy, among others are working on improving their efficiency and effectiveness.

What are Alternative Energies?

Alternative energy is any energy source that is an alternative to fossil fuel. These alternatives are intended to address concerns about such fossil fuels. Renewable energy is generally defined as energy that comes from resources which are naturally replenished on a human timescale such as sunlight, wind, rain, tides, waves, and geothermal heat.

Solar

In the Washington Post Vivek Wadhwa states that by 2020, solar energy will be price-competitive with energy generated from fossil fuels on an unsubsidized basis in most parts of the world. Within the next decade, it will cost a fraction of what fossil-fuel-based alternatives do.

The incumbent fossil fuel industry will be disrupted. The world will no longer depend on oil. There is a high probability that this will occur within 20 years.

Wind

Furthermore, Wadhwa points out that wind power, for example, has also come down sharply in price and is now competitive with the cost of new coal-burning power plants in the United States. It will, without doubt, give solar energy a run for its money. There will be breakthroughs in many different technologies, and these will accelerate overall progress.

There is little doubt according to Wadhwa that we are heading into an era of unlimited and almost free clean energy. This has profound implications.

The Renewables 2015 Global Status Report shows that once again more renewable energy has been generated this last year in comparison to the previous one. Over the course of ten years, the planet's renewable energy power capacity (excluding hydro) has grown from 85 to 657 gigawatts, which is almost an eightfold increase.

Going forward

The promise of sustainable energy that families can access directly will contribute to the disruption in this sector. Moreover, Musk's incipient success with Powerwall (and with Solarcity) can serve as an inspiration for more entrepreneurs to disrupt the energy sector.

10

Blockchain
BREAKTHROUGH TRANSPARENCY

*Technology made large populations possible; large
populations now make technology indispensable.*
—Joseph Krutch (Writer)

Think of a blockchain as a peer-to-peer database, an open source, massively distributed list, which can keep a record of pretty much anything. It is a decentralized ledger of transactions. The blockchain started as a public ledger of all Bitcoin transactions that have ever been executed. It is constantly growing as 'completed' blocks are added to it with a new set of recordings. The blocks are added to the blockchain in a linear, chronological order. However, it will not be limited to recording Bitcoin transactions; the blockchain technology is set to disrupt other realms.

According to BBC's Paul Coletti, other characteristics of the blockchain are: the blockchain is split up and distributed over thousands of computers all over the world; these computers compete to be the one to add the next block to the blockchain; clever cryptography ensures each block is digitally signed in such a way that changing or altering an entry invalidates every other entry preceding it going right the way back to block one—this is what makes each entry in the blockchain immutable, while the uncoordinated nature of the network is what keeps it honest. To corrupt the system you would have to commandeer more than 50 percent of the computers on the system.

What is Blockchain?

Blockchain is essentially a record, or ledger, of digital events—one that's "distributed," or shared between many different parties. It can only be updated by consensus of a majority of the participants in the system. And, once entered, information can never be erased. For example, the bitcoin blockchain contains a certain and verifiable record of every single bitcoin transaction ever made.

Investors like Marc Andreessen, Netscape cofounder, have poured tens of millions into the development and believe this is as important an opportunity as the creation of the Internet itself.

The blockchain is shared by all parties participating in an established, distributed network of computers. It records every transaction that occurs in the network, therefore eliminating the need for "trusted" third parties such as payment processors. This is why it has a revolutionary potential.

Blockchain advocates often describe this innovation as a "transfer of trust in a trustless world," because the entities participating in a transaction are not necessarily known to each other yet they exchange value with surety and no third-party validation. This is why the blockchain is a potential disruptor.

It is a secure network because each transaction is encrypted with a hash that is used to verify the succeeding hash. Simply put, this means that changing one vote requires millions of votes to be changed before another vote is cast. The network is protected by the simple fact that no hacker has enough computing power to rewrite so many votes that quickly. A hacker would actually need more computing power than the top 500 supercomputers combined, 256 times over!

Distributed cloud storage

The founder of future-of-cloud storage service Storj, Shawn Wilkinson says that with available excess hard drive space around the world, users could store the traditional cloud 300 times over. "Considering the world spends $22 billion plus on cloud storage alone, this could open a revenue stream for average users, while significantly reducing the cost to store data for companies and personal users."

Breakthrough with smart contracts

Innovating in this space is Ethereum, a platform for decentralized applications which allows developers to build blockchain applications aiming to "decentralize the Internet and return it to its democratic roots". It is a platform for building and running applications which do not need to rely on trust and cannot be controlled by any central authority. It is a decentralized platform that runs smart contracts: applications that run exactly as programmed without any possibility of downtime, censorship, fraud or third party interference. Proponents of smart contracts claim that many kinds of contractual clauses may thus be made partially or fully self-executing, self-enforcing, or both. Smart contracts aim to provide security superior to traditional contract law and to reduce other transaction costs associated with contracting.

End of patents

Blockchain technology could replace patents. A tech company might want to prove that it created a certain innovation on a specific point in time without filing for a patent. If anyone challenges ownership of a technology, the company could later reveal internal documents that are linked to the transaction hash, proving the existence of the innovation at a specific date specified on the Blockchain.

Electronic voting

Josh Blatchford of BTC.sx, a Bitcoin trading platform, says that the automation of counting paper votes is a no-brainer for cost, time, and accuracy improvements. However, previous systems have been riddled with technical issues. The main problems are the inability to verify a machine's accuracy during recounts and being prime targets for hackers. It is not surprising that political parties are turning to the blockchain for their internal voting.

On the accuracy front, blockchain's pseudonymity allows each vote to be publicly shared without identifying the voter. Hence, each voter could check their vote has been counted from public records. This may one day eliminate election corruption in the undeveloped world.

As Nic Cary of Blockchain.info said, '"Check it on the Blockchain" will be the phrase of the 21st century. It will be as commonplace as people saying "Google that."

New era of transparency?

The blockchain could enable accountability at an unprecedented level. It could spell the end of corruption by design. All movements of money would be tracked, and it would be easy to require those moving money to register their public keys, according to Bitcoin news journal, CryptoCoinsNews.

For example, Texas-based blockchain company Factom is hoping that this technology could help eliminate some of the corruption that has afflicted the property market in Honduras for years. Factom announced that the government of Honduras has agreed to partner with the company in order to create a new system to oversee the country's land title record keeping ledger. This could signal the beginning of the end for the old system allowing officials to hack into the records and assign prime real estate to themselves.

Similarly in the UK, former Member of Parliament, George Galloway wants to use the blockchain to keep track of the Mayor of London's spending.

We could be witnessing the start of a revolution that could signify the end of corruption by design, ushering in an era of irrevocable transparency leading to better democratic structures.

PART II

The Future of Work: Shifts Driven by Emerging Technologies

11

Reality blurred
VIRTUAL OR REAL?

Is this the real life, is this just fantasy? —Queen

Just as the Internet enabled the world to efficiently collaborate across countries, across organizations and time zones, virtual and augmented reality (VR, AR) are about to open a new realm for humanity: the ability to be somewhere else, without actually being there—that is the ability to be telepresent.

The implications for the future of work are significant. When VR and AR take off, the incentives to keep workers at home will outweigh the costs of providing staff with full-time offices, facilities and all the associated costs. It simply will not make business sense to have everyone physically working at the same place when circa 90 percent of office functions can be achieved in a virtual environment.

Virtual or real?

We will likely see a reversal in office attendance: instead of taking a day off to work from home, we will take a day off to go meet at the office. The office will become a hub to socialize "the old way" perhaps once a week, or once a month. The mention of "the office" will elicit the question: do you mean the virtual one or the physical one?

At the speed of development described by key players, working VR solutions for business should be available by 2020. By 2025 the norm will likely be working on a virtual environment, and physical (old-fashioned face to face meetings) will be reserved for very special occasions (which could nevertheless be achieved in a VR environment). This will disrupt the whole concept of working from home. It is hard to appreciate the power of VR as it has not gone mainstream just yet, but once we do, it will be hard to go back.

Disruption will be evident in many industries that rely on face to face interactions—which is basically every industry. We will need to reinvent ourselves for a new reality where the lines of virtual and real are blurred.

Even though today we see VR as cumbersome headsets, the miniaturization trend will also take over, so that common eye glasses and eventually contact lenses will serve as VR displays.

The main achievement of VR technology is creating *presence*. Creating the sensation of presence, *a la Matrix*, is within the realm of engineering achievement, and may be regarded as an achievement of remarkable importance, because of its implications.

Every industry will face some level of disruption due to this key technological achievement. Industries such as education, retail, health, and many others will be disrupted. Perhaps the most impacted at the beginning may be transportation and business travel.

The work commute in Mexico City

Take for instance a megalopolis of close to 22 million inhabitants, Mexico City. With millions of people commuting daily for hours, the impact of VR technology could be life-changing. VR could significantly curb the need to leave home for work. If everything that needed to be achieved in a physical work space could be achieved in a virtual environment, the need for transportation would be significantly curbed. This would have very important implications for a city as big as Mexico City, including less pollution, less street maintenance, less traffic, but most importantly, more time for everyone. On a monthly basis this could free up millions of man hours. On a conservative calculation of 5 million people commuting daily for an average of 2 hours, with a 20% adoption rate of VR, thus avoiding the daily commute, Mexico City could save one million man hours per day.

The future of working from home

Eventually, full-time employees will realize that they do not even have to live in the same city to go to work. They will be able to live increasingly farther and farther away from corporate headquarters, as the need to meet physically diminishes powered by VR tech improvements.

For many corporations, most recently Yahoo, working from home did not make the most sense, as collaboration was hindered.

In a VR context this is solved, as each worker is fully immersed and visible to each other, creating at least the same level of accountability and collaboration afforded by a traditional work environment.

However, innovations will be needed to optimize the way people work in a VR environment, so that we do not necessarily copy the old way of working in an office setting but adapt to the future of work in the virtual reality setting.

Innovations by Oculus Rift, HTC Valve, and many other VR players will enable this transformation. As we pointed out, Mark Zuckerberg has already "seen" this as the next big platform, creating an entirely new ecosystem. For many, this development could be as revolutionary as the Internet has been.

C-level executives will be able to live in small towns anywhere in the world, experiencing the quality of life they want. For most workers, there will be a requirement to collaborate in the same time zone, plus or minus 2 hours (if sleep is not to be disrupted). Otherwise workers will be free to live anywhere in the world where they can meet the connection specs. The societal impacts may be profound.

Business meetings

With virtual reality business meetings for instance, it is conceivable that the whole business travel industry, a €1 trillion global behemoth according to Forbes, could be at peril of being seriously disrupted. Why travel when you accomplish pretty much the same for the vast majority of meetings?

The new normal

The impact of virtual reality on work will be significant. The whole point of "going to work" will take on a whole new meeting. Soon enough actually being somewhere and doing it virtually will be indistinguishable. The need to meet people face to face will be seen as a luxury, as we will be able to collaborate more effectively in virtual environments.

Dr. Luis Flores Castillo points out that this is already the case at CERN, where technology is leveraged to facilitate interactions with colleagues on the other side of the planet. "When connecting to a meeting, you can share your screen, paste URL's and email messages in the chat box on the fly, and anyone interested can immediately add those to their notes." Arguably this form of virtual interaction is the precursor to true VR ushering in a new era of collaboration.

12

Blackbox combinations
THE NEW *LEGO* BLOCKS

A pile of rocks ceases to be a rock when somebody contemplates it with the idea of a cathedral in mind. —Antoine de Saint-Exupéry

Uber's very existence could not be understood without key innovations such as smartphones and accompanying app ecosystems. This is the case for countless companies that have been born on the Internet, and on newer platforms.

Many of the emerging technologies that we have reviewed in this book will serve as the new platforms that will enable the launch of the next round of entrepreneurs.

In how many different ways could you combine autonomous vehicles, with 3D printing, with biotech, with alternative energies, with VR and AR, with blockchain, with everything else? The truth is that imagination—creativity—will become the main constraint.

With technological platforms advancing at breakneck speed, increasing their sophistication, they will serve as building blocks serving to enable increasingly powerful innovations.

Blackboxing

There is another important quality of these technologies to consider: they can be seen as black box technologies, that is, each of these building blocks is a self-contained component, which can be used by programmers without a detailed understanding of how it works. This is a constructivist approach to science and technology referenced by French philosopher, anthropologist and sociologist of science, Bruno Latour. He states that blackboxing is "the way scientific and technical work is made invisible by its own success. When a machine runs efficiently, when a matter of fact is settled, one need focus only on its inputs and outputs and not on its internal complexity. Thus, paradoxically, the more science and technology succeed the more opaque and obscure they become."

This is a very powerful property. Once you incorporate any of these technologies into your business model, as long as you hit the upgrade button, you will have the most recent version of that particular component, without necessarily having to worry about the complex inner workings of that particular technology.

However as an entrepreneur it is essential to quickly increase the level of sophistication, and build upon more complex ones.

Heart of innovation

Channing Robertson, one of the Stanford professors of Elizabeth Holmes, the founder and CEO of Theranos, recounts that Elizabeth came up with a patent for a wearable patch that could not only administer a drug, but monitor variables in the patient's blood to see if the therapy was having the desired effect, and adjust the dosage accordingly.

Robertson recounts: "I remember her saying, 'And we could put a cellphone chip on it, and it could telemeter out to the doctor or the patient what was going on. And I kind of kicked myself. I'd consulted in this area for 30 years, but I'd never said, here we make all these gizmos that measure, and all these systems that deliver, but I never brought the two together."

What Robertson describes is precisely at the heart of innovation: the ability to connect 1+1+1 to make 5 or 10.

New Lego blocks for entrepreneurs

The wide array of emerging technologies available today means that entrepreneurs can literally play with more blocks in increasingly creative ways.

The blackbox combinational nature of emerging technologies means that the sum of the components do not simply add in value, but more likely multiply, in ways that are often unanticipated.

13
Augmented humanity
THE AGE OF AMPLIFIED COGNITION

"New developments in machine intelligence will make us far, far smarter as a result, for everyone on the planet. It's because our smart phones are basically supercomputers."
—Eric Schmidt, Executive Chairman, Google, USA

If you think about it, one of the byproducts of the exponential growth of computing power is increased cognition: an increased capacity for humans to access information, to know, to achieve, to make decisions, and to accomplish more with less effort. This will be done through a combination of devices and technologies, including AI agents, AR devices, memory enhancements, better voice recognition and so forth. As a consequence, the bar will be raised for everyone. Today the modern knowledge worker is expected to be proficient in finding information and using office software, as the bare minimum technologies. But in the near future, we will see that the unassisted, or un-augmented worker, will simply not be able to operate anywhere near the level of her augmented peer.

Augmented humanity is a term that was coined in 2010 by Google Chairman Eric Schmidt. It defines the use of technology to both aid, and replace, human capability in a way that joins person and machine as one. Schmidt stated that devices connected to cloud supercomputers could give us powers or "senses" that we didn't know were possible. This could be something like Watson-as-a-Service. "Think of it as augmented humanity" he suggested.

Human enhancement

According to the Institute for Ethics and Emerging Technologies, human enhancement refers to any attempt to temporarily or permanently overcome the current limitations of the human body through natural or artificial means. The term is applied to the use of technological means to select or alter human characteristics and capacities, whether or not the alteration results in characteristics and capacities that lie beyond the existing human range.

Human enhancement refers to the general application of the convergence of nanotechnology, biotechnology, information technology and cognitive science to improve human performance. Human enhancement technologies are techniques that can be used not simply for treating illness and disability, but also for enhancing human characteristics and capacities.

Human augmentation is not new

According to microchip implant pioneer, Amal Graafstra, "it's a myth that human augmentation is anything new. Since the first humans picked up sticks and rocks and started using tools, we've been augmenting ourselves." It is just that the tools have been shrinking in size and are getting easier to use. "The common thread," Graafstra says "is transhumanism; to constantly and fundamentally transform the human condition."

Futurist Gerd Leonhard questions: "Which university professor would not want to have the world's knowledge available instantly in the lecture theater using a Wikipedia-app controlled via a contact lens or an unobtrusive brain-computer-interface? Which doctor would not want IBM's Watson Analytics VR-display to provide him with real-time medical information and thereby protect him from malpractice lawsuits?"

If we are already witnessing a reliance on our smartphones as memory augmentation devices (who still memorizes phone numbers?) then it is logical to think that subsequent steps towards augmentation will be welcomed as the new "normal".

Human and machine

In *Rise of Robots*, Martin Ford argues that it will not be a case of men vs. machines, but men working with machines doing the work that an unassisted human can only do poorly. This makes the case for human augmentation as a desirable "upgrade" that could increase the value of the worker.

More examples of augmentation

In Codegent, Beth Gladstone cites some examples of augmentation, including Edios Audio, a mask that allows you to "zoom in" on nearby conversations, gaining a momentary sense of enhanced selective hearing. Google Glass's wink activates a snapshot feature, allowing you for instance to take a picture of where you parked your car. Spotify can now listen to the user's body to measure heartrate and choose music based on mood. A tiny device called Reveal LINQ, which sends irregular heart rate data from patient to doctor, via a 3G box that lives under your bed.

Man-machine convergence

The creativity in terms of augmentation is unlikely to see any bounds. Coupled with progress in Internet ubiquity, AR, IoT, increased computational power, big data, AI, and biotech among other trends, in five years' time we could be looking at very powerful augmentation devices and approaches, some of which we haven't yet thought of.

As Gerd Leonhard puts it "just imagine a world where you simply cannot compete or even keep up without some kind of wearable augmented reality (AR) or virtual reality (VR) device, or without an

implant, or other mental or physical augmentations."
This world might be closer to us than anticipated.

Expanding the neocortex

Ray Kurzweil points out the current trend we see in
augmented human cognition will only heighten. At
present smartphones represent the way in which we
enhance our memory and our capacity to move about
the world. However, as we have seen in the previous
two parts, this is only the beginning. Soon, we will be
able to further enhance ourselves to the point—
Kurzweil posits—where we expand our neocortex
into the cloud. Just think what will happen when
instead of having to Google something, you just
know, by virtue of your brain being connected to the
Internet. Kurzweil calls this hybrid thinking.
Converging technologies will make us smarter and
more efficient in a variety of ways.

"And so, over the next few decades, we're going to do
it again," says Kurzweil. "We're going to again
expand our neocortex, only this time we won't be
limited by a fixed architecture of enclosure. It'll be
expanded without limit. That additional quantity will
again be the enabling factor for another qualitative
leap in culture and technology."

14

Towards full automation
REPLACING HUMAN LABOR

*All of the biggest technological inventions created by man—
the airplane, the automobile, the computer – says little about
his intelligence, but speaks volumes about his laziness.*
– Mark Kennedy (Author)

U p until the industrial revolution muscle power was limited to what animals and humans could provide. With the advent of the steam engine, the availability of physical power grew exponentially, marking an era of tremendous progress. This is referred to as the first machine age by Erik Brynjolfsson and Andrew McAfee in their *Second Machine Age* book.

If muscle power was essential to that age, brain power is the key for the second machine age. However this is not the *natural* brain power of humans, but that afforded by computers. It is the era we live in, where plentiful computing power—which continues to grow exponentially—multiplies the availability of cognitive power. Today computers are doing the jobs that were reserved for humans not long ago. This is powering a new era of automation.

Brynjolfsson and McAfee point out that it is the "exponential, digital, and combinational" nature of technology that underpins the powerful nature of the second machine age.

Not so special after all

Human cognition is losing its value. It is no longer unique. This is because the supply of computing power is increasing until it becomes ubiquitous and increasingly sophisticated. Overall the price of a floating operation per second is dropping. All this powers a cocktail of technologies that make us humans increasingly replaceable. Humans may not be so special after all. Computer code is replacing basic human cognitive functions.

One of the consequences of the technological progress is precisely our ability to replace human cognition with machine cognition.

As we saw in the first part of the book, processes which were limited to human labor are now being performed by computer code. Automation will continue to replace predictable and repetitive labor.

Point in case is made by Delfina Eberly, Director of Data Center Operations at Facebook. She comments on the data server to (human) admin ratio, pointing out that each Facebook data center operations staffer can manage at least 20,000 servers, and for some admins the number can be as high as 26,000 systems. To accomplish a decade ago, it would have probably taken a small army.

Automation is not a new thing

However, automation is not a new thing. In a Pew Research Center study, Jim Warren, the founder and chair of the First Conference on Computers, Freedom & Privacy, wrote that "Automation has been replacing human labor—and demolishing jobs—for decades, and will continue to do so. It creates far fewer jobs than it destroys, and the jobs it does create often—probably usually—require far more education, knowledge, understanding and skills than the jobs it destroys."

Rex Troumbley, researcher at the University of Hawaii at Manoa, wrote, "We can expect robots, artificial intelligences, and other artilects to increasingly displace human labor, especially in wealthy parts of the world. We may see the emergence of a new economy not based upon wage labor and could be realizing the benefits of full unemployment (getting rid of the need to work in order to survive)".

The reason why this will start in wealthier parts of the world is simple: a worker in the developed world is more expensive than her peer in a developing country. There is simply more incentive for a corporation to start an automation process where wages are higher. In the developing world, where labor intensity is still affordable it will take longer, but it will also come of age.

AI and robotics will continue to displace low level worker skills

According to specialists, our continuing failure to re-train under-skilled workers will exacerbate un- and underemployment. This is because advances in AI and robotics will require workers that are increasingly educated and specialized. Those who attain the required skill levels will find new opportunities while under-skilled workers will likely be left on the curb.

Rebecca Lieb, an industry analyst for the Altimeter Group and author, said, "Enterprises will require a highly educated, digital and data literate workforce, which does not bode well for blue-collar workers, or softer skill white-collar workers. Given trends in U.S. education, this could lead to high demand for engineers from foreign countries (as we've seen in the past) with advanced degrees in engineering, mathematics, etc., as institutions of higher learning in this country fail to produce enough graduates with the requisite skill sets."

The rise and fall of secretaries

It only takes a few episodes of *Mad Men* to appreciate the full glamor of the typical office set in the '60s and with it the reliance on secretaries. As the economy shifted from factories to offices, much of this setup carried right through the early 1990s. In much of the U.S., and globally, the secretary became a common job, in many cases it was the gatekeeper to power. However the rise of the personal computer overturned this; PCs were able to do more and more of the work which had been exclusive to secretaries.

Jobs ripe for disruption: 47 percent

Vivek Wadhwa is an American technology entrepreneur and academic. He is a fellow at the Stanford Rock Center for Corporate Governance. Wadhwa states that "not only will there be fewer jobs for people doing manual work; the jobs of knowledge workers will also be replaced by computers. Almost every industry and profession will be impacted."

A 2013 study by Carl Benedikt Frey and Michael Osborne from Oxford University predicts that 47 percent of all U.S. jobs are under threat of automation. But as, Wadhwa states, the difference this time is that the threat goes beyond manufacturing jobs or those at the lower-wage end of the spectrum. This time automation impacts a wide spectrum of office jobs as well.

The Oxford study quantified the vulnerability of jobs listed by the U.S. Department of Labor. The researchers compared the detailed tasks of 702 positions in relation to the predicted future ability of technologies, specifically connected to advances in machine learning and robotics, concluding that almost half are under threat.

Jobs you would expect to be safe

Daniel Tencer of the Huffington Post Canada identified jobs that most people would not think would be threatened by automation. Some of these are:

Construction workers

Today developers can build entire components of buildings in a factory before shipping them to the construction site. That makes it much easier to hire robots to do the work, and automation of construction is spreading rapidly, especially in Japan. Probability of automation, according to the Oxford report: 88 percent.

Bakers

Automated bakery lines for food retailers are coming. Probability of automation, according to Oxford report: 89 percent

Journalists

The next wave of automation will take aim at the creative side of the business: reporting. The Associated Press this year announced it will use a bot to produce 4,400 articles on corporate earnings every year. Pretty soon we can expect sports stories, weather stories and local crime reports to be written by bots as well. Probability of automation, according to Oxford report: 55 percent (for editors)

Farmhands

Two pieces of technology will revolutionize farming in the twenty-first century: self-driving tractors and drones. Drones will soon be fertilizing and inspecting crops while self-driving tractors will pick crops. Farmers themselves though are not under threat of replacement (less than one percent chance). All the same, fully automated farms are now a thing. Probability of automation, according to Oxford report: 97 percent

Paralegals and legal assistants

Preparing for a big trial can take thousands of hours of poring through documents, in the discovery process, work that used to be done by paralegals and associates. Now all that time-consuming work is being done much faster by data-mining algorithms. Probability of automation, according to Oxford report: 98 percent (legal secretaries), 94 percent (paralegals)

Pharmacy workers

Automation will take over the counting and dispensing of pills, reportedly making far less dispensing mistakes than human pharmacists. Though pharmacists themselves will continue to exist, their technicians and aides could soon be out of work. Probability of automation, according to Oxford report: 92 percent (pharmacy technicians), 72 percent (pharmacy aides)

Hospital and medical workers

Doctors will carry diagnostic tools with them that will be instantly connected to hospital records and other databases. Essentially, the hospital will come to you. But that means fewer jobs for orderlies, administrators and other staff. This shift is probably still a few generations away. Probability of automation, according to Oxford report: 91 percent (medical records technicians), 90 percent (medical lab technologists), 47 percent (medical appliance technicians)

Real estate agents and related jobs

In the age of home sale sites like Zillow and Trulia, homebuyers can find out at least as much about homes available for sale online as they could from a realtor. The role of realtors is shrinking, and many other real estate-related activities are also becoming automated, such as property appraisal. Forbes contributor, Bruce Kasanoff recently suggested that 95 percent of a real estate broker's job can be better done by machinery. Probability of automation, according to Oxford report: 97 percent (real estate brokers), 90 percent (appraisers and assessors of real estate), 86 percent (real estate agents)

Airport security and customs officers

Automated self-service security systems could make today's human driven airport security a thing of the past. California startup firm Qylur is selling automated security booths that can detect suspicious-looking objects and sniff out dangerous chemicals. Similarly passport-reading machines are eliminating the need for a portion of the country's customs officers. Thomas Frey, a prominent futurist and head of the Da Vinci Institute, figures about 90 percent of airport security jobs will be automated within a decade. Probability of automation, according to Oxford report: N/A (84 percent for security guards).

It is clear that this wave would target jobs in the heartland of the middle and upper-middle class: professional occupations.

From being decision-support systems to decision-taking systems

As the ability of the machine to turn raw data into information and then insight improves, the space remaining for a human to add value shrinks and eventually disappears.

In this vision, a requirement for creativity is not necessarily a defense against automation. Computers can already write sports articles for newspapers which readers cannot distinguish from pieces penned by humans. A computer system called Lamus in Malaga, Spain, produces such sports columns.

Customer Service disrupted

A senior policy adviser for a major U.S. Internet service provider commented, "Virtually all customer service work involving telephonic and online contact with human beings will be rendered unnecessary by better communications and computing services and by AI. Vast amounts of manufacturing, maintenance and other lower-skill jobs will give way to robots. The nation will have failed to make the necessary changes in either its education system or its commitment to promote economic and social equality, so the impacts on those with lesser skills, training, and motivation will be dramatic, with some socially disruptive results."

AI poised to take over jobs

Artificial Intelligence is poised to take over a number of jobs, because it makes economic sense. We will go through times of trouble as we adjust to this new reality. It is already a humbling experience because many incredibly bright and talented individuals currently find in this market that their skills are not in demand.

The transition has started

The recent crisis of 2008 was perhaps our last big warning sign. After this we have seen that companies are not willing to be exposed to risks, especially with big employee bases. They will not go back to making massive hires anymore. The era of the huge bloated corporation is long gone. Actually this was gone a decade ago, when tech companies started to exhibit exorbitant ratios of company valuation to number of employees. Point in case, today Uber has a market cap of $50 billion and an employee base of around 2,500 (est.) giving it a ridiculously high valuation to employee ratio of $20 million. Google's ratio is $7.7 million per employee ($438.1 billion market cap: 57,148 employees). This was close to impossible in the old economy.

Automated trains in London?

On July 9, 2015, London was paralyzed. The Tube Union went on strike bringing the super convenient Underground service to a halt. In response to this the Mayor of London pointed out that, in line with automation trends, driverless trains are pretty much inevitable. Is the future catching up with us quicker than anticipated?

This book was not written by a robot

This book was not written by a robot (I promise). But a decade from now, I am certain it could well be. Sooner rather than later, I will have to accept that as a human writer I may not be as special as I thought I was. For humanity, it is a sobering realization—to think that a machine or an algorithm can replace our work.

We have seen the global shifts that replaced animal and human muscle power, mechanizing many functions, so we thought that jobs that had to do with brain power were safe. But this is not so anymore.

Jobs in high demand

When asked which types of profession might be in highest demand, MIT professor Erik Brynjolfsson states that, in the new economy, specialists, the creative class, and people who have jobs that require emotional intelligence, and people who create everything from writing and art to new products, platforms and services, will likely see higher demand.

Mark J. Schmit, PhD, executive director of the Society of the Human Resource Management Foundation states that jobs in health care, personal services, and other areas that are tough to automate will also remain in demand, as will trade skills and science, technology and mathematics (STEM) skills. Workers will need to engage in lifelong education to remain on top of how job and career trends are shifting to remain viable in an ever-changing workplace, said Schmit.

We will explore this in further detail in Part III.

PART III

Getting Ready for Disruption: Preparing for Exponential Technological Change

15

Education for disruptors
BEYOND THE INDUSTRIAL AGE

"What I really want out of life is to discover something new,
something that mankind didn't know was possible to do"
—Elizabeth Holmes, youngest self-made billionaire, founder
of Theranos, blood testing

I f we accept that the new knowledge worker is an augmented human capable of leveraging knowledge and emerging technologies to achieve what a small army of non-augmented humans could do a few years ago, then we have to seriously ponder: how do you teach a learner like that?

No more carbon copies

In an Industrial economy, education was designed to replicate workers, so that they were interchangeable pieces of well-oiled machinery. However, exact copies of worker are no longer as useful or relevant, because, by virtue of automation we see that eventually most predictable patterns will be ultimately replaced. Emerging education must recognize that learners in the new economy are moving towards an era of specialization, where workers and entrepreneurs will be highly rewarded for coming up with unique solutions.

Education will then move away from the mass-production of graduates towards highly customized educational programs. Instead of following a cookie-cutter approach to teaching and learning we will realize that it makes more sense to follow highly personalized teaching-learning methodologies which are adapted to each learner. While technology will serve as a key enabler of this, the biggest challenge will not be technological or even methodological, but cultural. We need to reconsider the role of education for the era we have entered.

Realization of the *student-worker-preneur*

One of the questionable assumptions relates to how the educational system sees the learner: is she an employee? Is she an entrepreneur? Is she a perennial student?

John Baker, founder of Desire2Learn, asserts that "life in the industrial economy was typically viewed as a series of discrete segments: school, work and retirement. But this thinking is no longer viable as we have entered the era of lifelong learning."

Are we then teaching students to be employees and not entrepreneurs? While we may be tempted to answer that everyone must be trained as an entrepreneur, it does not mean that everyone wants to be *exclusively* one or the other. The reality might lie somewhere in between: we need for learners to become proactive lifelong learners, who will likely work for a company as a full-time employee at some stage, and will more-than-likely start their own company, or be a freelancer. Hence a more balanced term which reconciles reality and work trends might be summarized in the realization of the *student-worker-preneur*, a term I have coined to represent that each of us is a student who is a worker and an entrepreneur in different degrees throughout our careers.

Memory augmentation: commoditized knowledge

If, for all practical purposes, knowledge is a Google search away, memorizing things will become irrelevant. The idea of regurgitating dates and names for the sake of it will be seen as a waste of time. Access to information will become increasingly commoditized and it will also be enhanced and sophisticated: from voice commands, to augmented reality displays, to automatic face recognition—the trend in memory augmentation is clear: we will need to memorize less and less. This implies that as educators the emphasis should *not* be placed on getting students to remember and regurgitate data. The case is strengthened by the increasing volume and speed at which information is generated; the body of knowledge in any given profession can change not in a matter of years but months or weeks.

Hence, knowing is not enough. The actual competence, doing, achieving something, is the real test.

Questioning the purpose of education

How can education keep up in times of exponential change?

Whereas in an Industrial age the quantity of graduates was the key variable to optimize, in the new economy, it will be the *quality* of graduates. Thus, instead of graduating professionals with the same (commoditized) skills, the most valuable education will be that which is able to cultivate the uniqueness of each learner, including an optimal mix of hard and soft skills, that is, technical and interpersonal competencies.

Learning to learn

Decades ago it used to be enough to learn a trade in a four or five year university program. However today, by some estimates, half of the technical information that you learn in a university program might be outdated by the time you finish.

In an era where new industries and business models are born overnight, it is clear that being able to learn at a record speed will not only give the learner a competitive advantage but it will become an essential skill for life.

However, as explored, the limitation for this is no longer access to information. Nowadays anyone can learn virtually any trade online, thanks to Massively Open Online Courses (MOOC), or through full university courses made available by Universities including Harvard, Stanford, and the Massachusetts Institute of Technology. While certification is still closely held by universities, the actual knowledge to be learned has increasingly become commoditized. As the new currency is being able to do, and not just knowing, the ability to proactively engage in self-taught education that helps develop real world competencies will become paramount.

Raison d'être: the motivation to learn

Underlying the ability to learn is the motivation to learn. In an article by the Center for Teaching and Learning at Stanford University, author Barbara McCombs, director of the Human Motivation, Learning, and Development Center at the University of Denver, is quoted on seven qualities of students who are optimally motivated to learn. McCombs points out that optimally motivated students see schooling and education as personally relevant to their interests and goals; they believe that they possess the skills and competencies to successfully accomplish these learning goals; they see themselves as responsible agents in the definition and accomplishment of personal goals; they understand the higher level thinking and self-regulation skills that lead to goal attainment; they call into play processes for effectively and efficiently encoding, processing, and recalling information; they control emotions and moods that can facilitate or interfere with learning and motivation, and; they produce the performance outcomes that signal successful goal attainment.

From my own experience working with students and business clients over the years, I see a clear correlation between the *motivation* for learning and the ability to learn. I would argue that it is more important to have a *reason* for learning, a powerful *why* that inspires the learner to pursue education.

This can be tied back to the importance of solving problems. To paraphrase McCombs, learning can be enhanced when the learner sees that what they learn can serve as a tool to impact the world in areas that are relevant to their own interests.

This is perhaps one of the greatest opportunities we have today: helping learners discover a reason and purpose for learning.

Passion, Curiosity, Imagination, Critical Thinking, and Grit

Peter Diamandis often gets asked a question about raising children in times of exponential change. "So, Peter, what will you teach your kids given this explosion of exponential technologies?"

"In the near term (this next decade) the lingua franca is coding and machine learning. Any kid graduating college with these skills today can get a job. But this too, will be disrupted in the near future by AI. Long-term, it is passion, curiosity, imagination, critical thinking, and grit."

Passion

"You'd be amazed at how many people don't have a mission in life. A calling, something to jolt them out of bed every morning," writes Diamandis.

Developing a passion is a key. It can be understood as the driving force, the true motivation behind work or any other endeavor.

"The best moments in our lives are not the passive, receptive, relaxing times," says Mihaly Csikszentmihalyi, author of *Flow,* "the best moments usually occur if a person's body or mind is stretched to its limits in a voluntary effort to accomplish something difficult and worthwhile."

In this state of flow a student-worker-preneur can be completely absorbed in an activity, especially one involving creativity. During this "optimal experience" you feel strong, alert, in effortless control, unselfconscious, and at the peak of your abilities, according to the author. The key to this is setting challenges that are neither too demanding nor too simple for a person's abilities.

In a talk at Singularity University, Ray Kurzweil, Google Director of Engineering, was asked "When robots are everywhere, what will humans be good for?" His answer was that, if under the logic that automation will take away a big chunk of the drudgery, the work humans don't enjoy doing, it will leave us with more time to explore what we want to explore. Part of his advice then was to "develop a passion."

American astrophysicist, cosmologist, author, and science communicator Neil deGrasse Tyson says that "what you need, above all else, is a love for your subject, whatever it is. You've got to be so deeply in love with your subject that when curve balls are thrown, when hurdles are put in place, you've got the energy to overcome them."

Developing a passion is closely linked to three other ingredients: curiosity, imagination, and critical thinking.

Curiosity

Jeff Bezos said this about success and innovation: "If you want to invent, if you want to do any innovation, anything new, you're going to have failures because you need to experiment. I think the amount of useful invention you do is directly proportional to the number of experiments you can run per week per month per year."

At an award's acceptance speech in London, Google co-founder Larry Page, said "we tried a lot of things, most of which failed." He elaborated that when they set out to create the world's biggest search engine, they were just pursuing their interests, hopefully arriving at something that would be useful. The key takeaway comes in the form of direct advice from Page: "You should pick areas that you think are interesting, that could be valuable, or where there's a lot of activity. I was interested in (URL) links because I knew no one else was interested in them, and I figured you could probably do something with them." We can infer from this that curiosity is key to arriving at what actually interests you.

The author of *Silicon Guild*, Peter Sims, points out the work of INSEAD business school professors who surveyed over 3,000 executives and interviewed 500 people who had either started innovative companies or invented new products. They concluded that a number of the innovative entrepreneurs learned to follow their curiosity. Without curiosity it would be impossible to expand the frontiers of what is possible.

Videogame inventor Will Wright, co-founder of Maxis (which became part of Electronic Arts) points out the importance of the joy of discovery: "It's all about learning on your terms, rather than a teacher explaining stuff to you." SimCity, one of Wright's creations, is an example of this.

Curiosity and the joy of discovery are closely linked to imagination, another quality identified by Diamandis.

Imagination

"Entrepreneurs and visionaries imagine the world (and the future) they want to live in, and then they create it. Kids happen to be some of the most imaginative humans around... it is critical that they know how important and liberating imagination can be," says Diamandis.

"Imagination is one of humanity's greatest qualities," says Richard Branson, founder of Virgin, "without it, there would be no innovation, advancement or technology, and the world would be a very dull place."

Critical thinking

"Critical thinking is probably the hardest lesson to teach kids. It takes time and experience, and you have to reinforce habits like investigation, curiosity, skepticism, and so on", says Diamandis.

A movement called *Philosophy for Children*, also known as P4C and under the auspices of Stanford University, began with the 1969 novel titled *Harry Stottlemeier's Discovery* written by the late philosopher Matthew Lipman. The novel and accompanying teacher manual were designed to help children in K-12 learn how to think for themselves.

Dr. Peter Facione, who spearheaded the American Philosophical Association's international study to define critical thinking elaborates on the meaning and importance of critical thinking: "We understand critical thinking to be purposeful, self-regulatory judgment which results in interpretation, analysis, evaluation, and inference, as well as explanation of the evidential, conceptual, methodological, criteriological, or contextual considerations upon which that judgment is based.... The ideal critical thinking is habitually inquisitive, well-informed, trustful of reason, open-minded, flexible, fair-minded in evaluation, honest in facing personal biases, prudent in making judgments, willing to reconsider, clear about issues, orderly in complex matters, diligent in seeking relevant information, reasonable in the selection of criteria, focused in inquiry, and persistent in seeking results which are fthe subject and the circumstances of inquiry".

Ad Astra school: "to the stars"

Speaking of education for disruptors, it makes sense to examine how disruptors are teaching their own kids. Elon Musk's disruptive endeavors span finance (PayPal), solar energy (Solar City), cars (Tesla), space exploration (SpaceX) and now, education. He didn't like his kids' school, so he started his own. It is called Ad Astra which means "to the stars". For now the school is also serving kids of SpaceX employees. One of its features is a focus on problem solving.

"Let's say you're trying to teach people about how engines work," said Musk to a media outlet. "A more traditional approach would be saying 'We're going to teach all about screwdrivers and wrenches'. This is a very difficult way to do it. A much better way would be, like, 'Here's the engine. Now let's take it apart. How are we going to take it apart? Oh, you need a screwdriver'." This is clear approach to ignite motivation and critical thinking. "It makes more sense to cater the education to match their aptitude and abilities," also remarked Musk. Interestingly, Musk reports that his kids "really love going to school" so much that "they actually think vacations are too long; they want to go back to school."

The Montessori approach

In a Wall Street Journal article, Peter Sims points out that "the Montessori educational approach might be the surest route to joining the creative elite." He cites that it is so overrepresented by the school's alumni that one might suspect a Montessori Mafia. Graduates include Google's founders Larry Page and Sergey Brin, Amazon's Jeff Bezos, and Wikipedia founder Jimmy Wales.

In an interview with Barbara Walters, Larry Page said: "we both went to Montessori school, and I think it was part of that training, of not following rules and orders and being self-motivated, questioning what's going on in the world, doing things a little bit differently."

114

The Montessori learning method was founded by Maria Montessori and it features a collaborative environment without grades or tests, multi-aged classrooms, as well as self-directed learning and discovery for long blocks of time, primarily for young children between the ages of two and a half and seven.

The approach nurtures creativity, taking after the work of inventors who typically improvise, experiment, fail, and retest. Sims points out that inventors such as Henry Ford and Thomas Edison were voracious inquisitive learners.

In a world flooded with often-conflicting ideas, baseless claims, misleading headlines, negative news and misinformation, you have to think critically to find the signal in the noise, explains Diamandis.

Grit

Finally, grit is seen as "passion and perseverance in pursuit of long-term goals," and it has recently been widely acknowledged as one of the most important predictors of and contributors to success.

Pinterest was launched in 2010. The story of co-founder Ben Silbermann is a great testament of perseverance. In 2008, Silbermann decided to quit a job he hated. However, he didn't know what he wanted to build, so built an app called Tote... and it flopped. He then decided to try a new idea, a site for collecting things, and it was rejected by many investors. He made fifty different versions of the site, launched it and got 200 initial users. Silbermann personally wrote welcome emails to his first 7,000 users, and in this process he discovered that his early adopters were "moms". The rest, as they say, is history. Today Pinterest is home to over 500 employees. The company recently doubled its valuation to over $11 billion.

Education and life as process of self-directed learning

Sergey Brin said "there are many important things to life aside from financial or career success, and in fact, it's not necessarily the ultimate success that motivates you, it's the process of getting there; the technology, the products that you build. I am not too concerned about finding something to do, though I do think it will be based on doing things that I really enjoy, and not have some end goal in mind."

Being exposed to new people and ideas

Speaking of predictors of career success, according to Ron Burt, one of the world's top network scientists, being in an open network instead of a closed one is the best predictor of career success, a discovery based on multiple, peer-reviewed studies.

Burt explained that if you are a member of a "large, open network where you are the link between people from different clusters", as opposed to being a member of a "small, closed network where you are connected to people who already know each other" you have a much higher chance of overall career success

"The more you repeatedly hear the same ideas, which reaffirm what you already believe. The further you go toward an open network, the more you're exposed to new ideas." Simmons concludes, based on network science, that people who are members of open networks, and hence open to all sorts of new information, are significantly more successful than members of small, closed networks.

The relevance of Science, Technology, Engineering, Math (STEM) education

The other key distinction in terms of education directly correlates with the first part of the book: emerging technologies.

The fact that a number of highly disruptive technologies are coming of age in a relatively short time frame presents an opportunity for student-worker-preneurs focused on Science, Technology, Engineering, Math (STEM). This is due to the competitive advantage that comes from being the first movers in those particular technologies.

Software guru, Jesse Stay, comments that, "there will be a much stronger, and greater need for engineering, and STEM-related jobs."

Overall employment trends by the US Labour Market Statistics, point out that graduates of Science, Technology, Engineering, and Mathematics (STEM) majors are and will be the most demanded areas. In the United States, STEM employment grew three times more than non-STEM employment over the last twelve years, and is expected to grow twice as fast by 2018.

Emerging technology companies will demand specialists in the areas we have reviewed, including 3D printing, advanced robotics, big data, biotech, nanotech, etc. presenting an economic opportunity of close to $20 trillion in the next 10 years. This will require specialized graduates in a wide array of industries, according to McKinsey & Co. However, as reported by Manpower and various studies, even at present, tech companies are struggling to find qualified candidates, resulting in unfilled positions and reduced growth.

Importance of the Soft skills: the 4 C's

Perhaps some of the hardest skills to teach, the so called soft skills, may be the most important ones in a new economy. While we have already referenced creativity and critical thinking, communication and collaboration will also be essential enablers for the modern student-worker-preneur. Referred to by some educators as the 4 C's, these soft skills are already instrumental in the workplace.

Leaders and Entrepreneurs

At the intersection of the technical and interpersonal competencies we can appreciate that two traits emerge: leadership and entrepreneurship. Arguably this is the intended result of the educational system. Moreover, I conclude that a focus on developing leaders and entrepreneurs might be the right educational aim, as this in consonance with the workplace shifts occurring over the next two decades, where less repetitive and predictable tasks are performed and where higher order tasks, in terms of cognitive complexity, will be the norm.

One of the projects we have started at Emtechub is precisely to identify young talented individuals from around the world who are doing impressive work with emerging technologies. They are emerging as leaders in their fields, addressing real world problems. We call it the Emerging Technology Leaders Global Initiative. Emtechleaders (for short) is a non-profit initiative that will help inspire young students around the world to pursue STEM careers, with a focus on emerging technologies.

Inspiring the young and young at heart

Neil deGrasse Tyson affirms that "Once you have an innovation culture, even those who are not scientists or engineers, poets, actors, journalists, they, as communities, embrace the meaning of what it is to be scientifically literate. They embrace the concept of an innovation culture. They vote in ways that promote it. They don't fight science and they don't fight technology."

Putting it all together

The Berkeley Alumni magazine points out that the inventor of the CRISP-cas9 DNA editing method, Jennifer Doudna "came to UC Berkeley from Yale in 2002 with a reputation for working side-by-side with Nobel laureates and having a knack for building alliances with other creative thinkers. She was also known for her brilliance at teasing out the purpose of biomolecules and for an uncanny ability to glean the shapes of the virtually invisible: the remarkable molecular machinery that spins within living cells".

This is a very telling statement. It not only reveals the importance of the hard technical skills, but how important it is to be able to collaborate, and to think creatively.

It strengthens the idea that the way forward in education has to do with a mix of hard and soft skills.

High tech companies are not only looking for proficiency in the hard, technical side of technology, but on the soft skills. In a Forbes article, Rich Milgram, CEO of career network Beyond, is quoted saying, "And more about how you think systems through and work within the context of the team. Learning a technology is the easy part. Having the mindset to apply it, having the mindset and logic to process it, being thorough and detail-oriented while doing so, these are the critical skills."

16

Solving problems
OPPORTUNITIES FOR ENTREPRENEURS

"Strive not to be a success, but rather to be of value"
—Albert Einstein

O ver the coming decades, we may just have to reinvent ourselves more than once. For example, if you currently give advice to clients, it is not far-fetched to envision an automated assistant that is able to understand natural language. It would be able to recommend solutions based on the experience and success rates of millions of users, carefully filtered by criteria to suit a specific client. This task is currently being automated.

Being unique

The answer then hinges around being unique and irreplaceable. Neil deGrasse Tyson says this: "I have a personal philosophy in life: If somebody else can do something that I'm doing, they should do it. And what I want to do is find things that would represent a unique contribution to the world - the contribution that only I, and my portfolio of talents, can make happen. Those are my priorities in life." This is quite a useful approach.

Greatest opportunity

Perhaps the greatest opportunity that emerging technologies gives individuals and organizations is the opportunity to be unique by finding unique ways to solve problems. The technologies we have explored represent a democratized approach to the solution side of the equation. What we must think of is what to solve for.

Today a team of entrepreneurs can single-handedly tackle highly relevant problems. The insurmountable is now within the realm of possibility for more individuals and organizations.

Focus on problem solving

One of the defining traits of entrepreneurs is a dogged focus on solving highly relevant problems. The problem with first picking a tool, i.e. a technology, and then finding a problem to solve, is that if you pick a hammer every problem will likely seem like a nail.

It would make more sense to first understand what exactly you are trying to solve, or what function you are trying to optimize.

In a business world where the actual day to day work becomes less and less relevant it will be more important for a person to focus on actually solving problems and challenges in innovative ways.

Abilities like problem-solving, entrepreneurship, creativity, communication and collaboration skills, as we have explored in the previous part, will become highly sought after. This would be one of the last skillsets to be automated.

Getting comfortable with emerging technologies

It goes without saying that those with at least some awareness of emerging technologies will be in a better position to solve problems. However, as mentioned previously, the black box nature of technologies also means that you do not necessarily have to understand the complexities of a technology to be able to use it, just as you do not need to understand the inner workings of the A380 to fly between continents. It is however useful for the creative process to become as familiarized as possible with all the different blocks that you could play with, so that you are able to create more and more complex creations.

Working with Machines

The other relevant insight would be the importance of mastering technologies that would suppose a short-term advantage in the market.

For example, just as anyone today is expected to know how to use a word processor or email, it does not mean that everyone masters these technologies. Hence workers could in theory gain an advantage by being early adopters of the emerging technologies we have reviewed.

Disrupt

A more proactive approach, of course, is to not sit around waiting for disruption, but to be the disruptor. An example of this is Elon Musk, PayPal co-founder and the real life Iron Man. In 2001, he was certain that NASA had plans to go to Mars, but when Musk searched NASA's Web site for details on this, he found nothing. "I thought there was some kind of mistake," Musk says. "I expected to find that they were well on their way and that we'd have to figure out something else to do. But there was nothing at all," Musk told Esquire. This stirred Musk to ultimately start SpaceX, risking his personal proceeds from PayPal. He is now in the midst of reinventing space exploration.

Space exploration is one of those realms that until recently was a country-level initiative, beyond the reach of individuals. But then came Elon Musk, one of the most daring and disruptive entrepreneurs on the planet.

Instead of basking in the sun to enjoy the healthy proceeds of his PayPal exit, Elon Musk decided to tackle some of humanity's most pressing challenges.

One of these has to do with the future of humankind outside of Planet Earth. He has made it his mission to help our species colonize Mars.

We may still be two, three or more decades away from a successful human colony in Mars, but think about this: within your lifetime there is a good chance that people will be living on the surface of another planet. How crazy is that?

17

Engineering socio-economic security

THE POST-JOB WORLD

Perfection is achieved, not when there is nothing more to add, but when there is nothing left to take away. —Antoine de Saint-Exupéry

A question at the heart of the future of work has to do with social and political responsibility: how can we ensure that we win *together*? We can be certain that machines and algorithms will continue to take over repetitive and predictable jobs currently performed by humans. Many white-collar jobs considered "safe" today will be gone in ten years. While the march towards full automation is generally unchallenged, two relevant questions remain: *how fast* will this happen and if the amount of jobs made redundant will be significant enough to upend the economy and ultimately trigger a collapse?

While the 'disrupt or be disrupted' approach may serve to drive innovation, another layer of innovation is essential: at the socio-political level.

Simply put, within one to two decades we may realize that there are not enough jobs for everyone, that the job deficit was never surmounted, but deepened.

New industries will appear, but they will not be as labor-intensive

While the top ten in-demand jobs in 2010 did not exist in 2004, we may see that by virtue of automation, the new types of jobs created in the future of work may not be as labor intensive. Furthermore "no matter what the jobs of the future are, they will surely require greater skill and education," says Vivek Wadhwa. This benefits those with higher levels of education and skill, but may completely exclude those that cannot qualify. It already is the case.

Jobs as a mechanism to distribute wealth

In economic terms, jobs are the main vehicle for distributing wealth in society. But this vehicle is quickly losing its relevance both in quantity and quality.

In the *Second Machine Age* book Erik Brynjolfsson and Andrew McAfee remind us that "in today's capitalist economies, most people acquire money to buy things by offering their labor to the economy. Most of us are laborers, not owners of capital."

However global economic trends are clear in that the share of wages are becoming a smaller share of national income, whereas the share of capital is increasing. In proportional terms more income is going to owners and less is going to workers.

Perhaps the most critical piece of evidence, according to Brynjolfsson, is a chart that depicts productivity and total employment in the United States. For years the two lines closely tracked each other, with increases in jobs corresponding to increases in productivity, as businesses generated more value from their workers, the country as a whole became richer, which fueled more economic activity and created even more jobs. But, beginning in 2000, the lines diverged; productivity continued to rise, but employment suddenly withered. By 2011, it was evident that despite economic growth there was no corresponding increase in job creation. Authors Brynjolfsson and McAfee refer to this as the "great decoupling", positing that technology has enabled both the vigorous growth in productivity but a weak growth in jobs.

In a presidential address, Irish economist Paul Sweeney expressed that "over three decades, the share of national income going to labor in most countries has been in decline. Conversely, capital's share of national income has increased. The trend in the decline in labor's share of income has been less noticeable as national income has continued to rise. This trend may have major consequences for economies, with reduced demand and also for societies, where social cohesion may be threatened by the trend... [Factors] include technology and increased returns to capital; globalization; the reduction of labor's bargaining power." Who has this hit the most profoundly? The lower skilled sectors, those with lowest educational levels.

Sweeney argues that the main impact of less job opportunities for the unskilled is growing inequality, which may threaten social cohesion, if median incomes remain stagnant while those at the top continue to rise. He argues that the decline in labor share has to become a mainstream policy issue for governments.

Martin Ford, author of *Rise of the Robots* concurs. He points out that despite an increase in productivity thanks to information technologies, the purchasing power of workers has eroded. He argues that we will reach a point where jobs are so specialized that fewer and fewer will qualify. This is because increasingly, not only entry level jobs but white collar ones will be taken over by algorithms. This has created a widening wealth gap, with profits concentrated with capital owners.

With exponential technologies, including IT, this gap is only getting wider. Will we reach a point when this is no longer sustainable? It leaves us in a predicament. Company owners will thrive in this new economy where a limited augmented human workforce can accomplish the work that required five times the amount of labor in the past. But the social cost will be high. For some, the scenario of unemployment can rise to a level that threatens social peace.

Day of reckoning

If the jobs requiring lower skill continue to be automated, and soon after white collar jobs face the same luck, we will be left with only a handful of highly specialized ones. This trend may never be reversed.

In one or two decades, after automation has made many positions redundant, not everyone will be able to take up a highly specialized job.

Jim Warren, in a Pew Internet Research study argues that "It is becoming more and more obvious that we (all developed nations) need to move—rapidly!—away from work-based income and well-being, towards and into more humane cultures, where all citizens are assured a comfortable quality of life, even if they aren't (and cannot become) sufficiently competent and expert to fill the shrinking number of more-demanding jobs that are available now, and will be still fewer in the future."

Guaranteed Basic Income

One of the models that is currently being explored is Guaranteed Basic Income. It is the proposal of a guaranteed baseline of income that covers the cost of living, including housing, rent, and the basic necessities. In Maslow's terms, this is the base of the pyramid.

From a capitalist perspective it makes sense if we consider that an economy without consumers is not really an economy. Furthermore widespread unemployment could threaten social stability.

Within ten years we will finally accept that there is a growing portion of the population which is highly educated, but still unemployable. This will be a structural issue.

Making a basic income work for everyone

In *Rise of the Robots*, Martin Ford says that a basic income should be tied to measures such as gaining education, performing community service, or participating in environmental projects. This might motivate people to work instead of spending all of their time in holographic worlds. However this might complicate it with additional layers of bureaucracy, deciding what is right for individuals.

Vivek Wadhwa writes in the Washington Post that "If we can develop the economic structures necessary to distribute the prosperity we are creating, most people will no longer have to work to sustain themselves. They will be free to pursue other creative endeavors. The problem, however, is that without jobs, they will not have the dignity, social engagement, and sense of fulfillment that comes from work. The life, liberty and pursuit of happiness that the constitution entitles us to won't be through labor, it will have to be through other means."

A pragmatic approach

What if the prospect of a 47 per cent unemployment rate materialized? It could spell economic collapse. Instead of speculating what could happen, perhaps the most sensible approach is to actively start experimenting with pilot programs pioneered by governments around the world. The aim would be to find modalities of a basic guaranteed income that could both ensure the basic needs of every citizen while providing incentives for improving valuable life and work skills.

Once again let us not be fooled by our linear intuitions into thinking that this problem is too far away. While no solution may be perfect at first, it is not too soon to start debating and piloting solutions that may become useful with increasing waves of automation.

18

The transition to prosperity
GATEWAY TO A BETTER WORLD?

The years comprehended between 2015 and 2025 will serve as a transition period, where we will begin to fully appreciate the pros and cons of automation and the wider array of emerging technologies. While we can be optimistic about the prospects of the future of work, this period may feel uncertain.

We will need at least two decades to fully transition to an era where needs are met more effortlessly and where poverty is eradicated and significantly greater social justice is achieved, enabled in part by technology.

Abundance and possibility

Let's imagine a scenario where freight costs lower thanks to 3D printing, transportation costs decrease due to autonomous vehicles, cheaper energy becomes available, as 3D printed houses become more available. Thanks to increased transparency the public cost of corruption could significantly diminish. With less need to travel for work owed to VR and AR the cost of doing business could decrease.

Overall, with less human inputs required, the cost of products and services attributed to wages could be dramatically slashed. All of this could spell an economy where needs could be met with less effort, at lower costs. This is a plausible scenario powered by emerging technologies.

The role of humans

As more and more repetitive tasks are automated we will transition to an economy where human labor is increasingly specialized and unique. This would presumably give way to roles that provide a higher degree of satisfaction, even though jobs would be scarcer.

Meaning outside of work

In regards to the sense of meaning and pride attached to working, society at large may have to come to terms with finding meaning outside of work. "The role of humans will be to direct and guide the algorithms as they attempt to achieve the objectives that they are given," says Martin Ford.

"No matter what fresh insights computers unearth, only human managers can decide the essential questions, such as which critical business problems a company is really trying to solve. Just as human colleagues need regular reviews and assessments, so these 'brilliant machines' and their works will also need to be regularly evaluated, refined—and, who knows, perhaps even fired or told to pursue entirely different paths—by [human] executives with experience, judgment, and domain expertise," says a McKinsey & Co. report.

Less drudgery, and more leisure time

Paul Jones, a professor at the University of North Carolina and founder of ibiblio.org, responded, "I for one welcome my new robot masters. I don't welcome the loss of jobs or the depersonalization of services. The social impact is and will continue to force us to refocus on what makes us human, who we are in relation to each other, and the terms of the social contract that binds us. In the South we saw great changes when the plantation system was abandoned. Not for the best—much room for improvement—but certainly for the better."

A professor at Stanford Law School wrote, "Robotics and similar technologies will displace lots of jobs. But those people will find productive things to do— not necessarily in fields created by robotics, but with the time that these advances have given them. Robotics and self-driving cars will free up substantial parts of our day. For some that will be a pure benefit; for others it will be partial compensation for the loss of work."

What makes us human anyways?

Gerd Leonhard expresses the view that in the future of work and jobs we might be moving towards the right-brain, become more human and increasingly less like machines. Ironically, this is completely the opposite of what traditional education may look like, avoiding emotions, limiting imagination, and sticking to schedules and plans. If you believe that non-algorithmic, i.e. emotional or subconscious factors such as trust, purpose, ethics, and values will remain at the core of human societies in the foreseeable future, this will clearly put a much stronger emphasis on the right brain. Education, training and learning will be changed forever as a consequence and we are already seeing the tip of that iceberg emerging.

Conclusions and final thoughts
THE CHOICES AHEAD

The human spirit must prevail over technology.
—Albert Einstein

When technological progress was fairly linear, say in the 1960's, it was relatively easy to anticipate demand and the range of changes that might occur in a given short and medium term horizon. Today we see that within five years a single company can achieve a valuation of half of the global market (Uber), and that another one (Airbnb) can be worth more than a chain of hotels established five decades earlier (Hyatt).

This is the true power of exponential technological progress, enabling disruption at a speed which was simply impossible some decades ago. If we also take into account the combinational nature of the tools we will have at our disposal, it is hard to predict exactly what will happen in five, never mind ten, years from now—we live indeed in exciting times of possibility.

We can see that greater abundance will be within reach, as the price of the factors of production drop. We are also transitioning to an economic model not based on scarcity but on abundance.

We will soon enter an era where our own creations match the computational power of our own brains, something which was only in fantasy land for most two or three decades ago.

But we must make some decisions both at the career, organizational, and socio-political level.

Harnessing disruption

The world of work as we know it, and the jobs we expect, will continue to shift, with greater speed. New industries will continue to rise seriously disrupting others. We are at an exciting time in history, with technological innovations looking more like magic, to paraphrase Arthur C. Clarke.

The relevant question is how does anyone become a player and not an idle bystander, or worse, an economic casualty of technologically-driven change? Perhaps the more urgent question from a social perspective is: are we heading to a future where only a few skilled ones benefit and mass inequality progresses?

One thing is clear: the better we understand the tools at our disposal, the better positioned we will be as individuals and as a collective to harness innovation to our advantage.

The fast fish eats the slow one

We may have to adapt at the individual level, at the organizational level, and at the socio-political level.

The speed of change means that thinking on your feet and being completely adaptive to models is the new currency. The founder of the World Economic Forum, Dr. Klaus Schwab, states that the motto affirming that the big fish eats the small fish may be outdated. Today it might be more appropriate to consider the *fast* fish eating the slow one—if not ask Kodak, which went bankrupt after failing to capitalize on the rise of digital photography. Kodak moved too slowly too late.

Choice

We now live in a time of exciting entrepreneurship marked by a sense of possibility. Perhaps this can be leveraged to lift more and more people out of poverty, and other conditions which may be deemed unacceptable in a world of such technological prowess.

We enter an era with many challenges related to the future of work, especially for those with limited access to education. However it is also an era of optimism, of increasing possibility to achieve greater equality and access to wealth and justice.

Technological advancement is not deterministic. We are not destined to do any one thing or the other. We, the humans, have the choice. Perhaps that is one of our defining features.

Glass box effect

"If you want to keep a secret, you must also hide it from yourself."—George Orwell, 1984

One of the effects of emerging technologies is that our lives and the lives of organizations and government will become increasingly transparent. This could be a *glass box effect*. Given the projected advances of the Internet of Things, including the increasing amount of information we will be submitting via wearable devices strapped on to monitor our health for instance, as well as advancing smartphone technology, we can expect an explosion of data that will reveal more and more information about individuals and organizations. Just as we feed data in our social media interaction, the incentives to volunteer increasing amounts of information will outweigh the potential privacy concerns, but nevertheless we will feed this information about ourselves.

For instance, we will start seeing insurance companies offering to lower your fees based on dynamic risk assessment. This could include your cholesterol levels, how fast you normally drive, the activities you perform on a day to day basis, etc. While we might see this as an invasion of privacy and opt-out, it would make sense for a lot of users who lead a low risk lifestyle and therefore justify participating. Those who opt out by default would pay a heavier prime.

However it is important to note that transparency does not mean loss of security. It means that the parties that should be privy to our information will receive it. Increasingly we will find the convenience of this, in many cases. This will push for greater transparency in institutions as well.

Bye, bye, privacy

Consciously or unconsciously we are making a decision that feeding our platforms with information is beneficial to us. And in most cases it is.

The stated goal of search engines is not to give you the information that you need when you need it, but to *anticipate* your desires. This will be accomplished thanks to the vast amount of decisions, likes, searches, etc. that each of us makes and has made in the past, which are documented.

This could turn out to be a double-edged sword. The real nightmare in the next decades will be escaping the reach of this know-it-all computing conglomerate which could eventually transition into the superintelligence.

The trust revolution

While the glass box effect may be seemed as a negative outcome enabled by emerging technologies, it could also represent a new era of transparency that could significantly empower new levels of structural trust—in the corporate world and especially in politics. Marc Benioff, Salesforce Chairman, states that "trust is a serious problem, we have to get to a new level of transparency—only through radical transparency will we get to radical new levels of trust."

However the other side of transparency is the right for privacy by individuals and organizations. "We've got to get this balance between privacy and at the same time use of data for legitimate public safety. The Internet is one of the greatest global goods, and common goods. If we destroy it, we destroy a lot of our economic future," says Microsoft's Satya Nadella.

Tim Berners-Lee, Professor of Engineering, MIT Computer Science and Artificial Intelligence Laboratory, expresses his hope that the discussion around tech-related privacy is more optimistic. "I hope that if we come back here next year we'll be talking about things very much more positively. So rather than just worrying about the niggling fears that people are going to be spying on us and doing the wrong things with our data, instead I hope we will have woken up to the completely exciting possibilities when it comes to our data."

Water is filling up the lake

Our world is shifting. The water is filling up the lake. Remember that in the year and half, two years, before it fills completely it will be only half full, and before that it will be only a quarter of the way. The temptation to dismiss exponential trends will be strong. Let's get ready for disruption. However, while it is tempting to approach this with a sense of pessimism fueled by the potential fallouts of the very technologies we are creating, it is the very technology we create which leads to possibility.

For young students

My final word for young students and entrepreneurs is that you find in this fascinating world of science and emerging technologies your own world, one that interests you, one that captivates you. But most importantly perhaps, I urge you to find problems to solve, ways to improve this world. As you can see sprinkled throughout the book and, especially in the world around, there are many challenges to solve.

The world around you allows you to solve countless problems, that is, once you determine what you are solving for.

Leaders and entrepreneurs needed

As previously mentioned, my conclusion is that a focus on developing leaders and entrepreneurs is highly relevant, given the work shifts that are projected in the coming decades. With the automation of drudgery, jobs that require a higher degree of cognitive complexity—including both hard and soft skills—will be more prevalent. This calls for an emphasis on education that develops the student-worker-preneur to fully realize personal and professional potential.

From dreams to reality

"The Internet was also once a dream, and so were computers and smartphones. The future is coming and we have a chance to build it together."-Mark Zuckerberg

Larry Page urged young graduates at a London gathering, saying: "work really hard on the things that really matter in the world and don't be afraid to fail." Albert Einstein famously said: "Strive not to be a success, but rather to be of value".

I leave you with this note written two decades ago by a 9-year old girl to her father:

"Dear Daddy, what I really want out of life is to discover something new, something that mankind didn't know was possible to do."

She is Elizabeth Holmes, Founder of Theranos, invited by the White House to be a Presidential Ambassador for Global Entrepreneurship, focusing on attracting women to science, tech, engineering and math fields.

I sincerely hope that this work on emerging technologies and the future of work may inspire you to continue your path of innovation.

Keeping in touch

INVITATION TO REGISTER

You can register to receive *Disruption* updates at www.emtechub.com/disruption

Follow us on Twitter: @emtechub and @vicdelr

Notes

Made in the USA
Middletown, DE
22 August 2017